Llyn y Parc
BETWS-Y-COED
River Conway
Llyn Elsi
Afon Machno
PENMACHNO

200
300
400
500
600
700
800
900

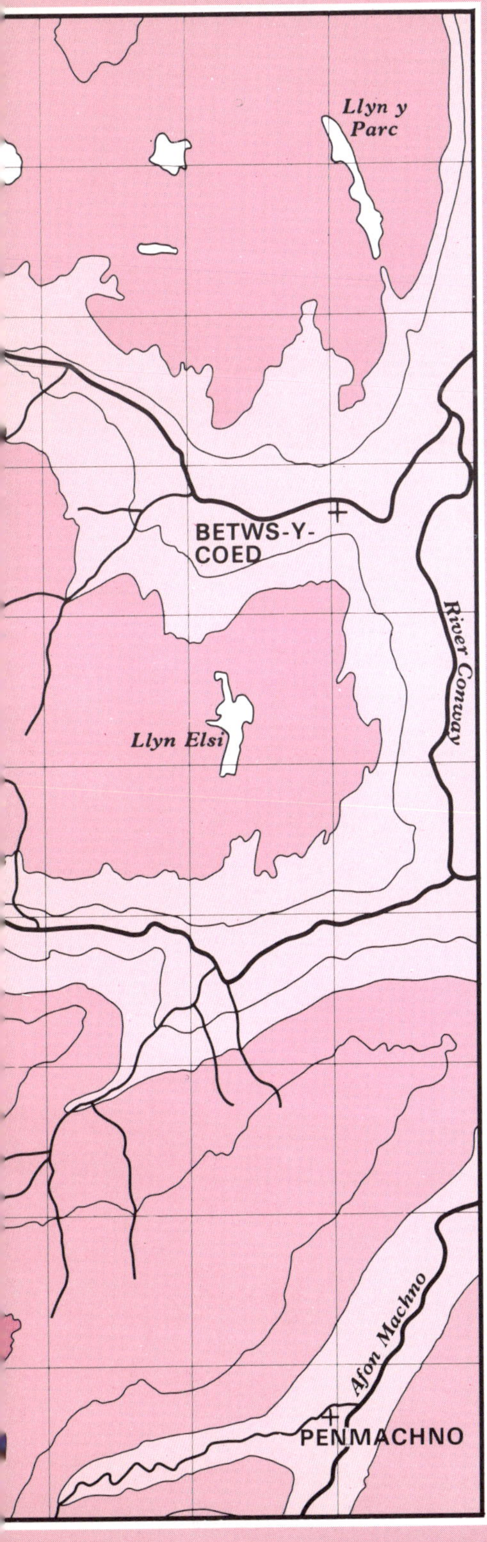

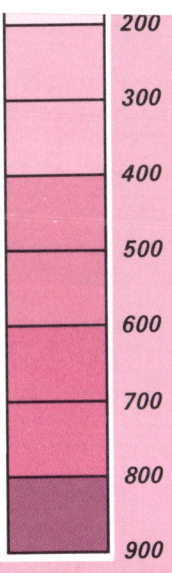

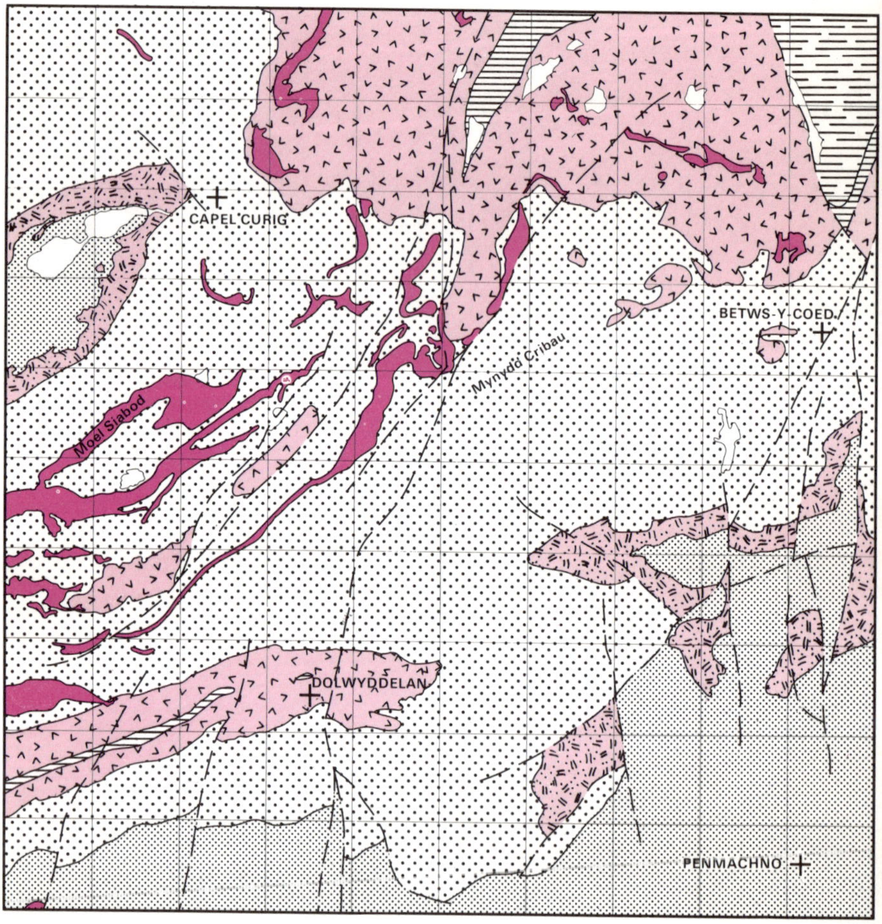

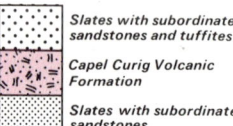

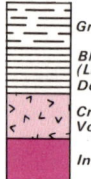

Figure 1
Geological sketch map of the area

Classical areas of British geology

M. F. Howells
E. H. Francis
B. E. Leveridge
C. D. R. Evans

Capel Curig and Betws-y-Coed

Description of 1:25 000 sheet SH 75

INSTITUTE OF GEOLOGICAL SCIENCES
Natural Environment Research Council

London   Her Majesty's Stationery Office   1978

© *Crown copyright 1978*

Bibliographical reference
Howells, M. F., Francis, E. H., Leveridge, B. E. and Evans, C. D. R. 1978 *Capel Curig and Betws-y-Coed: Description of 1:25 000 sheet SH 75*. Classical areas of British geology, Institute of Geological Sciences. (London: Her Majesty's Stationery Office.)

Notes

National Grid references are given in the form [7430 5001] throughout; all lie within the 100-km square SH.

Numbers preceded by E refer to thin sections in the collections of the Institute of Geological Sciences, and numbers preceded by RV or DT to specimens in the fossil collections.

Authors

M. F. HOWELLS, BSc, PhD
E. H. FRANCIS, DSc, FRSE
B. E. LEVERIDGE, BSc, PhD
C. D. R. EVANS, BSc, PhD
Institute of Geological Sciences,
Ring Road Halton, Leeds LS15 8TQ

Illustration credits

All illustrations are by IGS except the following: Brian P. Elkins, cover, figures 7, 10, 11, 17, 19, 26, and the small drawings on pp. 6, 10, 11, 13, 14, 16, 23, 26, 27, 30, 31, 35, 40, 45, 48, 50,

Forestry Commission, figures 9, 27, 28

Geological Society of America and University of Kansas, graptolite drawings on pp. 41, 43 from 'Treatise on Invertebrate Paleontology'.

ISBN 0 11 880783 8

Her Majesty's Stationery Office

Government Bookshops

49 High Holborn, London WC1V 6HB
13a Castle Street, Edinburgh EH2 3AR
41 The Hayes, Cardiff CF1 1JW
Brazennose Street, Manchester M60 8AS
Southey House, Wine Street, Bristol BS1 2BQ
258 Broad Street, Birmingham B1 2HE
80 Chichester Street, Belfast BT1 4JY

Government publications are also available through booksellers

Institute of Geological Sciences

Exhibition Road, London SW7 2DE
Murchison House, West Mains Road, Edinburgh EH9 3LA

The full range of Institute publications is displayed and sold at the Institute's Bookshop at the Geological Museum, Exhibition Road, London SW7 2DE

The Institute was formed by the incorporation of the Geological Survey of Great Britain and the Museum of Practical Geology with Overseas Geological Surveys and is a constituent body of the Natural Environment Research Council

Design by HMSO Graphic Design
Filmset by Jolly & Barber Ltd, Rugby
Illustration origination by Grantown Graphics Limited, London
Printed in England for Her Majesty's Stationery Office by Ebenezer Baylis and Son Limited,
The Trinity Press, Worcester and London

Dd 496301 K80

Preface

The first geological survey of the district by J. B. Jukes, W. T. Aveline and A. R. C. Selwyn was on the one-inch scale, published as an Old Series Sheet (78 SE) in 1852. The survey on which the present account is based was carried out by Drs Howells, Francis, Leveridge and Evans between 1968 and 1970 on the six-inch scale, supplemented extensively by aerial photographs. The published map (SH 75) covering the Capel Curig and Betws-y-Coed district is one of a series of sheets on the 1:25 000 scale being produced by the Institute of Geological Sciences to delineate the details of the complex Lower Palaeozoic geology of North Wales. The present account is designed to be read in conjunction with the map.

A.W. Woodland
Director

3 March 1977

Contents

1 Introduction 1
 Sedimentary rocks 3
 Volcanic rocks 6
 Tectonic setting 7

2 Carneddau Group 9
 Penamnen Tuffs and underlying strata 9
 Strata between the Penamnen Tuffs and the Capel Curig Volcanic Formation 10
 Capel Curig Volcanic Formation 11
 Strata between the Capel Curig Volcanic Formation and the Crafnant and Snowdon Volcanic groups 15

3 Crafnant and Snowdon Volcanic groups 21
 Crafnant Volcanic Group 22
 Snowdon Volcanic Group 33

4 Llanrhychwyn Slates and Black Slates of Dolwyddelan 41
 Llanrhychwyn Slates 41
 Black Slates of Dolwyddelan 42

5 Grinllwm Slates 44

6 Dolerites 45

7 Structure 48
 Folds 48
 Cleavage 49
 Faults 50
 Structural history 50

8 Mineralisation 53

9 Pleistocene and Recent deposits 56
 References 59
 Excursion itineraries 63
 Fossil localities 69
 Glossary 71
 Index 73

List of illustrations

1 Geological sketch map   iv
2 Generalised vertical section   2
3 Generalised ribbon diagram of the Carneddau Group and the Snowdon/Crafnant Volcanic Groups   4
4, 5, 6 Photomicrographs of the Garth Tuff and Lledr Tuffs   12
7 Sketch of the top of Curig Hill   17
8 Map and section of Curig Hill   19
9 The Llugwy Valley and Moel Siâbod   22
10 Massive tuff in the Lower Crafnant Volcanic Formation   24
11 Lower and Middle Crafnant scenery around Llyn Bodgynydd   25
12 Forestry road section in the Middle Crafnant Volcanic Formation   26
13 Depositional and post-depositional structures in the Middle Crafnant Volcanic Formation   28
14 Post-depositional structures in the Middle Crafnant Volcanic Formation   29
15 Photomicrograph of muddy crystal tuff from the Upper Crafnant Volcanic Formation   32
16 Pseudo-cleavage mullions in the Lower Rhyolitic Tuff Formation, Dolwyddelan   34
17 Moel Siâbod   35
18 Correlation of the Lower Crafnant Volcanic and Lower Rhyolitic Tuff Formations   37
19 Dolwyddelan Castle   38
20 Photomicrograph of basic pumiceous tuff from the Bedded Pyroclastic Formation   39
21 View from Moel Siâbod   44
22 Photomicrograph of dolerite with ophitic texture   46
23 Distribution and orientation of structures   49
24 Waterfall at Pont Cyfyng, Capel Curig   51
25 Abandoned mine near Sarnau   52
26 Lodes and properties in part of the Llanrwst mining field   53
27 Llyn Elsi, south-west of Betws-y-Coed   55
28 Swallow Falls   58
Plans of five walks   63–67
Caradoc shelly fossils   68

Introduction

This publication describes the geology of the district covered by the 1:25 000 map SH 75, comprising the country around the villages of Capel Curig, Betws-y-Coed, Penmachno and Dolwyddelan (Figure 1). Formerly in Caernarvonshire, the district now forms part of the county of Gwynedd and falls within the Snowdonia National Park. From elevations of less than 20 m above OD on the flood-plain of the Conway Valley in the east, the ground rises to 872 m at the summit of Moel Siâbod in the west. It is for the most part high moorland, much of it craggy, and is utilised mainly for sheep grazing and afforestation. Arable land is restricted to the valley floors flanking the rivers Llugwy, Lledr and Machno, all of which drain eastwards to join the River Conway.

Sedgwick made a traverse of the district from Nant Ffrancon through Moel Siâbod and Dolwyddelan to Penmachno in 1831 (Clark and Hughes, 1890, p. 520), but the first systematic investigation of the geology of the district was the primary one-inch survey begun in 1848 by Jukes, Aveline and Selwyn. Their Geological Survey maps and horizontal sections, published between 1851 and 1854, established the broad outlines of the stratigraphy and structure and were further illuminated by the two editions of the classic memoir on North Wales by Ramsay (1866, 1881). Supplementary contributions to the petrography and correlation of some of the volcanic rocks were made by Harker (1889) and Travis (1909). Three areas, together covering about one third of the district, were mapped on the six-inch scale for the first time by Williams (1922) around Capel Curig, by Williams and Bulman (1931) around Dolwyddelan and by Davies (1936) north of the Afon Llugwy. Some results of the recent primary six-inch survey by the Institute are embodied in papers by Howells, Leveridge and Evans (1973) and Francis and Howells (1973).

The sequence within the district is shown graphically in Figure 2 and is listed, with thicknesses generalised, on p. 2. Most of the named divisions are refinements of terminology first applied by Williams (1922), Williams and Bulman (1931) and Davies (1936). The exception is the Carneddau Group, which is here introduced as an alternative to 'Glanrafon Beds' —a term used in various senses by previous workers (p. 9). The group is here defined to include all strata from the base of the Ordovician to the base of the Crafnant and Snowdon

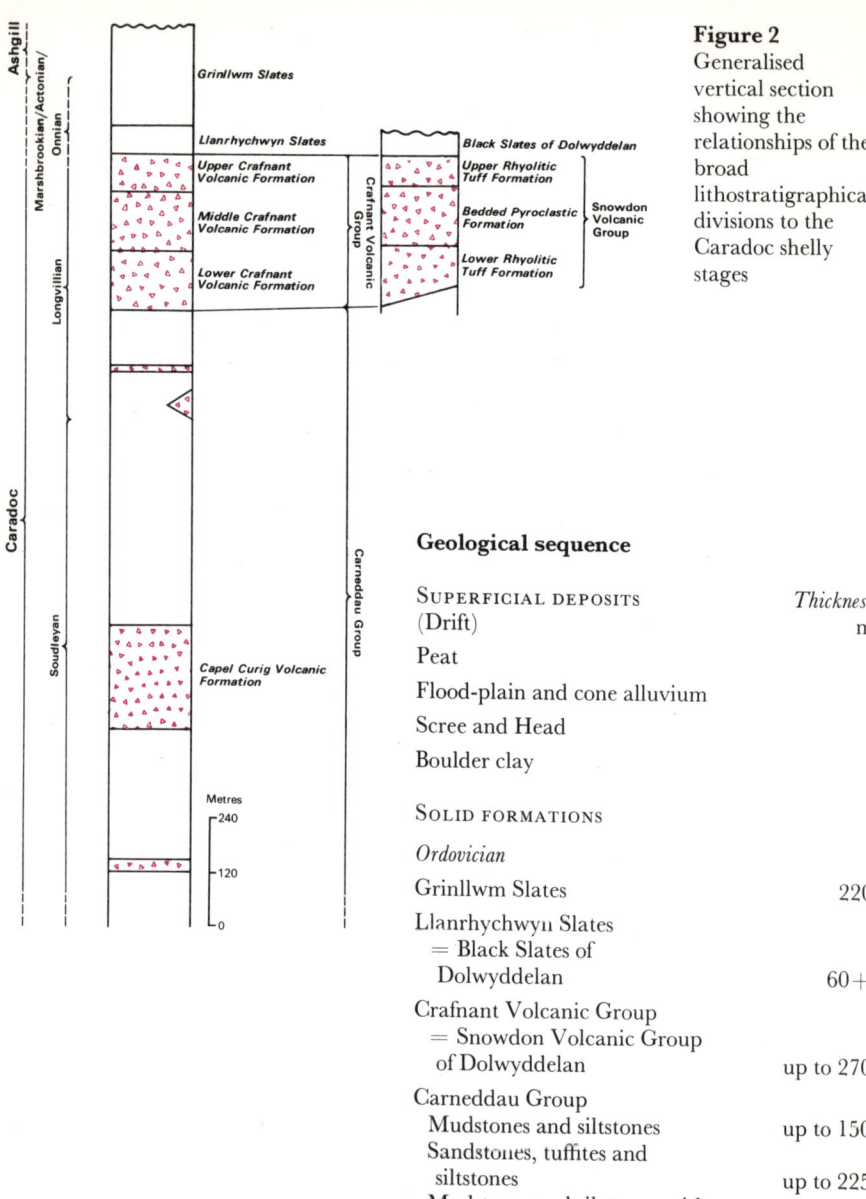

Figure 2
Generalised vertical section showing the relationships of the broad lithostratigraphical divisions to the Caradoc shelly stages

Geological sequence

	Thickness m
SUPERFICIAL DEPOSITS (Drift)	
Peat	
Flood-plain and cone alluvium	
Scree and Head	
Boulder clay	
SOLID FORMATIONS	
Ordovician	
Grinllwm Slates	220
Llanrhychwyn Slates = Black Slates of Dolwyddelan	60+
Crafnant Volcanic Group = Snowdon Volcanic Group of Dolwyddelan	up to 270
Carneddau Group	
Mudstones and siltstones	up to 150
Sandstones, tuffites and siltstones	up to 225
Mudstones and siltstones with subordinate sandstones	up to 600
Capel Curig Volcanic Formation	120 to 220
Mudstones and siltstones with subordinate sandstones	450
Penamnen Tuffs	30
Mudstones and siltstones	125+

Intrusive Igneous Rocks
Dolerites of Ordovician age

Volcanic groups. The lowest beds of the group do not crop out within the district.

Fossils occur so sporadically in this mixed sequence of volcanic and sedimentary rocks, that many problems of faunal correlation remain to be solved. Collections made during the survey (p. 69) complement earlier work summarised by Diggens and Romano (1968), Romano and Diggens (1969) and Bassett (1972), to suggest that the whole of the sequence exposed in the district falls within the Caradoc and Ashgill series of the Ordovician. The lowest assemblage so far found within the district is a Soudleyan fauna obtained from the Capel Curig Volcanic Formation. The Soudleyan–Longvillian boundary can now be drawn fairly confidently within the sandstone-tuffite-siltstone division in the upper part of the Carneddau Group (Figure 2). The change from shelly faunas to graptolite faunas in the upper part of the Snowdon/Crafnant Volcanic Group, however, lends uncertainty as to the position of the top of the Longvillian and the presence of the three highest stages of the Caradoc. For example, although Williams and Bulman (1931) assigned the Black Slates of Dolwyddelan to the upper part of the *Dicranograptus clingani* Zone, faunas recollected there and from the equivalent Llanrhychwyn Slates during the survey are referred to the underlying *Diplograptus multidens* Zone (p. 70). Moreover, there seems to be no agreement as to where the *D. multidens/D. clingani* zonal boundary should be placed relative to the stage boundaries (compare for example Whittington and Williams, 1964, table 1, with Williams, 1972, fig. 2). The overlying, apparently unfossiliferous, Grinllwm Slates are assigned, in part at least, to the Ashgill on the basis of their correlation with the Bodeidda Mudstones of Conway.

Sedimentary rocks

The sediments of the Carneddau Group are mudstones, siltstones and sandstones containing brachiopod and trilobite faunas. Except for the uppermost beds, they indicate a continuing fairly shallow marine environment in which sedimentation broadly kept pace with subsidence. The sandstones are of greywacke type; their lateral impersistence and repeated intercalation with siltstones and mudstones (Figure 3) may reflect local instability at the margin of a subsiding basin. Unpublished work by the Institute in the ground immediately to the north-west of the district suggests that the coarser terrigenes of the group are fluviodeltaic and neritic rather than turbiditic, with facies variations and current directions indicating derivation from a landmass to the north-west.

Towards the top of the Carneddau Group there is, throughout the district, an upward passage from siltstones to mudstones which indicates an increased depth of water. This increase is further apparent in the overlying Lower Rhyolitic

Figure 3
Generalised ribbon diagram showing lateral variations in the Carneddau Group and Snowdon/Crafnant Volcanic groups between Dolwyddelan and Betws-y-Coed. Intrusions omitted

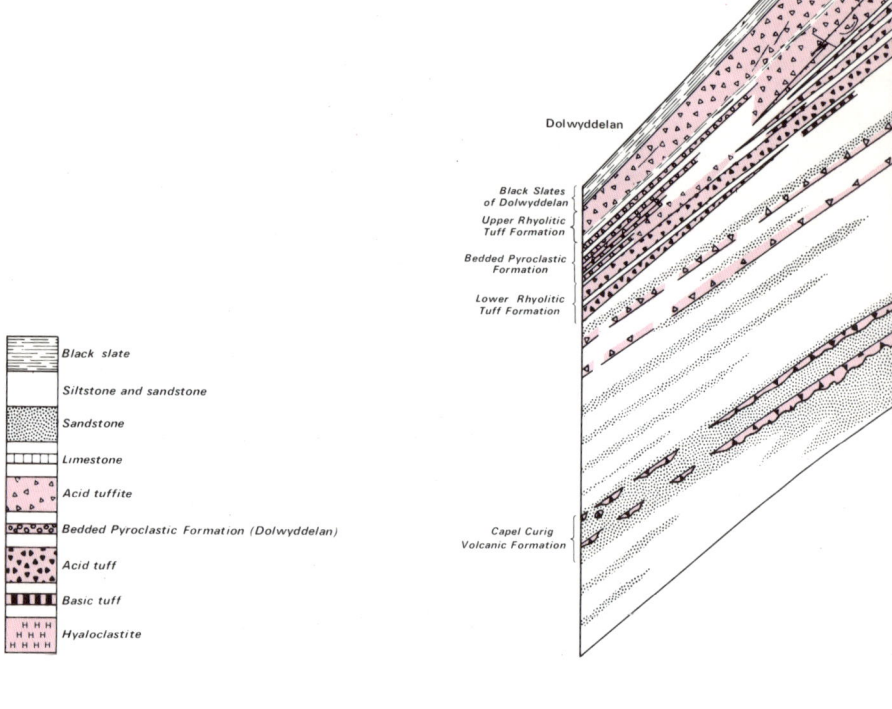

Tuff Formation and Lower Crafnant Volcanic Formation where, despite the rapid incursions of thick volcanic deposits, the intercalated sediments are of siltstone and argillaceous limestone rather than of sandstone. The lithologies and sedimentary structures in the Middle Crafnant Volcanic Formation are also entirely deep water in character. The penecontemporaneous accumulations of basic rocks in the Bedded Pyroclastic Formation caused local and temporary shallowing around Dolwyddelan. Towards the top of that formation there, and at the same horizon elsewhere in the district, the widespread establishment of a deeper water environment is signalled by the onset of a black graptolitic mudstone deposition which continued during and after the volcanic episode represented by the Upper Crafnant and Upper Rhyolitic Tuff formations.

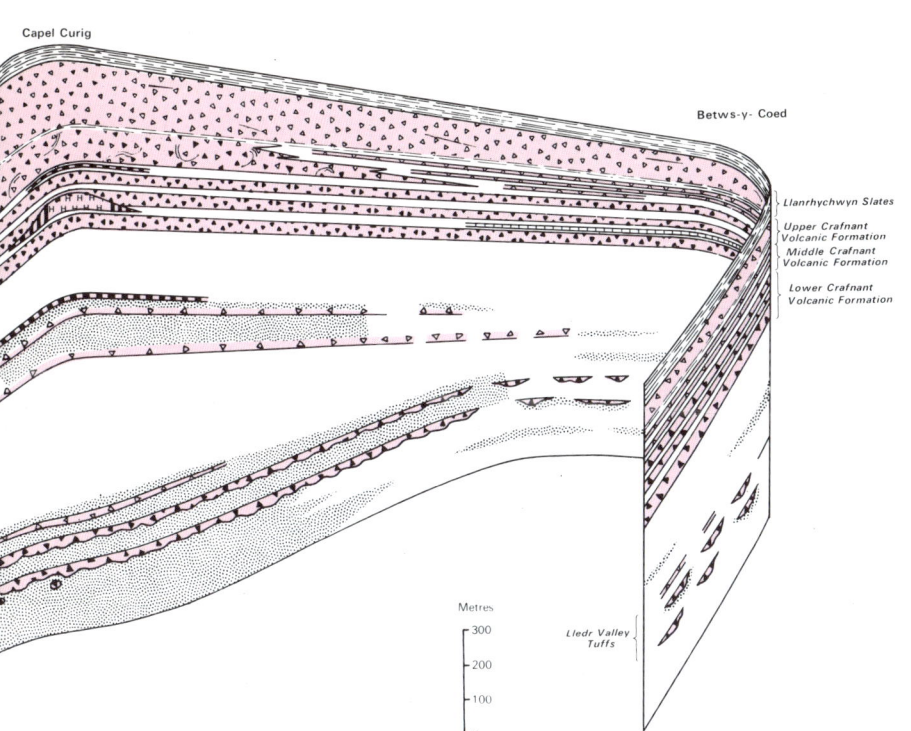

During the deposition of the Carneddau Group, while the environment was still one of shallow water, the pattern of sedimentation may have been influenced by contemporaneous volcanism in one of two ways. The first is illustrated by the Capel Curig Volcanic Formation. In the north-western part of the district there is a change from dominantly arenaceous sediments below the volcanic rocks to dominantly argillaceous above (Figure 3), which may reflect the emptying of magma reservoirs under volcanic centres farther north and north-west, so that an initial shallowing, represented by the sandstone immediately overlying the tuffs, was followed by an overall deepening of the depositional environment. The effect did not extend as far from the volcanic centres as the Dolwyddelan area, where alternations of sandstone and siltstone occur both below and above an attenuated sequence of tuffs.

The second type of modification of sedimentation by volcanism is inferred from the sandstone and tuffite division higher in the Carneddau Group. Here the upward sequence of siltstone, tuff or tuffite, sandstone is so commonly repeated as to suggest that the relatively rapid accumulation of pyroclastic material in the shallow-water environment temporarily altered the balance between deposition and subsidence. The suggestion is further supported by the reworking at the tops of those pyroclastic beds and their local impersistence.

Volcanic rocks

Large volumes of both basic and acid volcanic rocks occur within the sedimentary sequence. The basic rocks consist mainly of dolerite intrusions, but they also include tuffs and hyaloclastites which are of such limited lateral extent that there can be little doubt that they emanated from local, relatively short-lived, eruptive centres within the district. Their lithologies indicate that the eruptions were mainly submarine, though at late stages some volcanoes may have emerged temporarily above sea level before being rapidly eroded. Their possible relationship to the dolerite sills and the broader issue of the almost simultaneous availability of basic and acid magma fall outside the scope of this account. The geochemistry of the intrusions and basic extrusive rocks in the context of plate tectonics is the subject of a separate work (Floyd and others, 1976). Most of the volcanic rocks of the district, however, are acid tuffs, consisting of shards and crystals with a minor proportion of lithic clasts and matrix. They exhibit both welded and unwelded textures and are typically the products of ash flows (MacDonald, 1972). Associated with them are subordinate amounts of acid air-fall tuffs and tuffites. Many of the ash-flow tuffs cropping out around Capel Curig were formerly described (Williams, 1922; Davies, 1936) as rhyolites and their association with marine sediments led Williams (1922) to suppose that they were submarine flows. Following the recognition of welded shard textures first in the Ogwen area (Oliver, 1954), and subsequently elsewhere in North Wales (Rast and others, 1958; Beavon and others, 1961), many rocks previously called rhyolite were reclassified as ignimbrites. The reclassification was accompanied by a re-

Eutaxitic structure in welded tuff

Eruption in Snowdonia in Caradoc times

vision of the belief that they were submarine flows, for welding was equated categorically with subaerial environments of eruption and deposition—a concept reconciled with the stratigraphy by postulating repeated emergence of volcanic islands for ash-flow emplacement, followed by repeated subsidence for deposition of the intercalated marine sediments (Rast, 1961; Brenchley, 1964, 1969).

Results of the recent survey have cast doubt on the need to postulate subaerial environments either for welding or for ash-flow eruption and emplacement. In the Capel Curig Volcanic Formation there is a gradual upward transition from strongly welded massive tuff to current-bedded ripple-marked unwelded tuff. Contrasting this tuff lithology and the associated fossiliferous sediments with the lateral equivalent of the formation to the north and west of the district, Francis and Howells (1973) postulated a submarine environment of emplacement within the district and a subaerial environment of eruption and emplacement to the north and west. It is implicit in this interpretation that welding can occur in a submarine environment if the eruptive source is subaerial.

In the Lower Crafnant/Lower Rhyolitic Tuff Formation, the tuffs are unwelded, though composed dominantly of comagmatic material characteristic of ash flows. From their lithology and their associated sediments they have been interpreted as the products of submarine eruptions in which hot suspensive gas was gradually replaced by water to give turbid flows (Howells and others, 1973). Submarine eruption and emplacement is also invoked for the last major volcanic episode in the district — the Upper Crafnant/Upper Rhyolitic Tuff Formation. Graptolitic mudstones were being deposited in a deep-water environment that had been established before the episode began and the heterogeneous admixtures of ash-flow debris and unlithified muds which are characteristic of the formation, probably resulted from secondary slumping.

The inference to be drawn from the volcanic rocks of the district and their palaeogeographical setting is not that the eruptions and surfaces of emplacement were entirely submarine, for the inclusion of air-fall tuffs in various parts of the sequence and the evidence of the Capel Curig Volcanic Formation shows that at least some of the eruptions were subaerial at some stage. It is rather that the presence of ash flows in a marine sequence does not necessarily imply repeated regional or local emergence with concomitant unconformities and disconformities.

Tectonic setting

The district forms part of the Welsh Basin, in which thousands of metres of strata accumulated during Lower Palaeozoic times. The basin is generally assumed to have been separated from a proto-Atlantic ocean to the north by an Irish Sea landmass or horst which Dewey (1969) suggests was related to an earlier Benioff Zone. Evidence for the landmass is provided in Lleyn and Anglesey, where there is a marked overstep by the Ordovician (George, 1961, 1963) and indications of penecontemporaneous uplift and thrusting (Shackleton, 1954). The landmass is presumed to have been the source of the coarser clastics of the Carneddau Group.

The whole of the Ordovician volcanism in North Wales is interpreted by Fitton and Hughes (1970) as part of an island-arc system related to the plate tectonic model of Dewey (1969). Geochemical work on the dolerites, however, suggests to Floyd and others (1976) that continental calc-alkaline volcanism of 'Andean-type' is more likely than a calc-alkaline island-arc environment. Suggestions that the large volume of acid material was derived by crustal fusion have not been challenged, but it is not clear whether the surface expression of the acid volcanism was from fissures or central calderas: the only caldera so far postulated, in Central Snowdonia (Shackleton *in* Beavon, 1963; Rast, 1969; Bromley, 1969), by no means finds general acceptance (see for example Fitch *in* Bromley, 1969). Whether fissures or calderas, there were evidently several eruptive centres, only one of which – the source of the single rhyolite flow near the top of the Carneddau Group, near Betws-y-Coed – lay within the district. The source of the Capel Curig Volcanic Formation appears to have lain west to north-west of the district; the lower, main flows of the Lower Crafnant/Lower Rhyolitic Tuff formations came from the west, and the uppermost flow from the north.

The evidence suggests that the acid volcanism originally found subaerial expression in the uplifted area related to the Benioff Zone and that the centres became progressively submerged during Caradoc times as the depositional regime became one of deeper water.

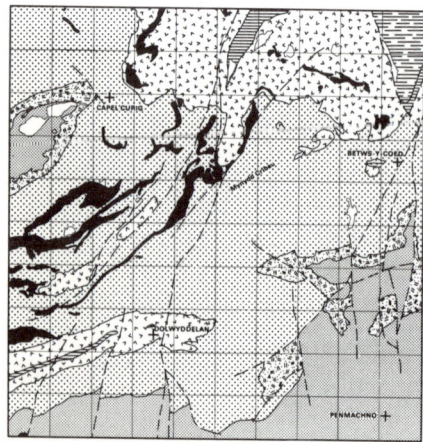

Carneddau Group 2

The literature dealing with the Ordovician strata below the Snowdon and Crafnant Volcanic groups contains inconsistencies in nomenclature, particularly in the various usages of 'Glanrafon'. The term was first used in Central Snowdonia by H. Williams (1927) who divided the beds between the 'Llanvirn (Maesgwm) Slates' and the Snowdon Volcanic Group into 'Llandeilo (Glanrafon) Slates' below and 'Gwastadnant Grits' above. These lithological divisions remain valid only as far north as Nant Ffrancon (D. Williams, 1930). To the south-west and south-east they give way to a laterally variable sequence of slates and sandstones with subordinate tuffs to which Williams and Bulman (1931) and Beavon (1963) applied the term 'Glanrafon Beds'. As H. Williams (1927), D. Williams (1930) and Shackleton (1959) recorded a lithological, as well as a faunal change at the top of the Maesgwm Slates, their 'Glanrafon Beds' meet the modern specifications of a lithostratigraphic division and is used in that sense (modified to 'Glanrafon Group') on the published 1:25 000 geological map of Central Snowdonia. It is clear from that map, however, that neither the Maesgwm Slates nor any equivalent Llanvirn strata have been recognised south of Snowdonia. Similarly, and more relevant to this account, no Maesgwm/Llanvirn line can be drawn farther north into the country between the Carneddau and the Conway Valley, where Davies (1936) and Stevenson (1971) have used 'Glanrafon Beds' in a more restricted sense than that applied in Central Snowdonia. Thus for this northern tract, of which the Capel Curig–Betws-y-Coed district forms part, the term Carneddau Group is proposed. The group is here defined (p. 1) to include all strata between the base of the Ordovician and the base of the Snowdon and Crafnant Volcanic groups.

Extent of the Carneddau Group

Penamnen Tuffs and underlying strata

Carneddau Group beds underlying the Penamnen Tuffs crop out mainly in the vicinity of Penmachno. They consist of up to 125 m of siltstone and slate, with ribs of false-bedded sandstone forming small exposures throughout the steep forest slope north of Afon Glasgwm.

The type section of the Penamnen Tuffs is located at the head of Cwm Penamnen, just outside the southern boundary of the district, where two units of acid tuffs, each up to 9 m thick, are separated by a similar thickness of siltstones. The outcrop extends into the district south-east of Pen-y-Benar, on the steep western side of Cwm Penamnen [7340 5001] where part of the higher tuff unit is seen in a small exposure. The tuff has a strong vitroclastic fabric of devitrified shards within a fine aggregate of chlorite and sericite, with crystals of sodic feldspar, a few crystals of quartz and isolated rounded lithic clasts, mainly of iron-rich siltstone.

Outcrops of tuff farther east, near Penmachno, are correlated with the Penamnen Tuffs. One can be traced for nearly 1 km through thick forest 1 km WNW of the village. The field relationships of two further isolated exposures around Bryn Bedyddfaen, to the north-east of the village, are uncertain. There is, however, a close lithological similarity between the rocks exposed around Penmachno and those of the type area. All are rich in shards, which range widely in shape and size and are moderately tightly packed. Crystals of sodic feldspar showing unusual resorbed peripheries are numerous (E 37352), quartz crystals are fewer and generally smaller. Lithic clasts, mainly of pumice and siltstone, are invariably present.

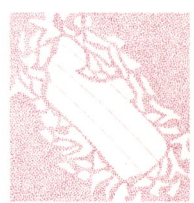
Resorbed sodic plagioclase

Strata between the Penamnen Tuffs and the Capel Curig Volcanic Formation

South of Dolwyddelan the Penamnen Tuffs are overlain by about 360 m of sediments composed predominantly of siltstone which are best exposed in the crags south of Clogwyn-y-Benar. They show an overall upward coarsening from mudstones at the base to siltstones with thin beds of sandstone showing ripple-drift and cross lamination at the top. The top of the sequence is ill defined in this area, for the Capel Curig Volcanic Formation consists of isolated masses of tuff, closely associated with coarse- to medium-grained sandstones. Farther east, however, the 350 m of strata between the Penamnen Tuffs at Penmachno and the more persistent Lledr Valley (Capel Curig) Tuffs of Rolwyd and Iwerddon consist of interbedded sandstones and siltstones together with one thin local tuff [776 515]. In the tight anticlinal structure in the Lledr Valley, the equivalent beds consist mainly of siltstone, but sandstone, generally fine-grained, occupies the core of the fold and is well exposed on the southern limb of the structure, upstream from Gethin's Bridge.

Only the upper part of the sequence is seen in the Capel Curig Anticline where 100 to 135 m of sandstones with rare impersistent intercalations of siltstone form the core of the structure. The sandstones, exposed principally on the steep slopes of Creigiau'r Garth, are greenish, cross-bedded and medium- to coarse-grained, with local pebbly lenses. They owe their colour to a high content of finely divided chloritic material (E 36831) which probably represents altered volcanic detritus.

Cross-bedding

Capel Curig Volcanic Formation

The formation crops out in four areas within the district. In three of these, namely in the type area around Llynnau Mymbyr, in the Lledr Valley and in isolated faulted ground north and west of Penmachno, it consists of two or three tuffs separated by sediments. In the fourth area, south of Dolwyddelan, the formation is represented by scattered isolated bodies of tuff which have ovoid outcrops. Correlation of individual tuffs from one area to another is uncertain, but the broad correlation is supported by their stratigraphy, their distinctive petrography and most importantly by their unusual relationships with subjacent sediments. At their lower contacts some of the tuffs locally transgress the underlying sediments at angles up to 90°, with minor apophyses resembling magmatic intrusions. Some of the apophyses are detached from the main mass of the tuffs so as to form discrete bodies simulating tuff-pipes at outcrop. The tuffs are generally welded and in most places the eutaxitic foliation remains parallel to the regional dip, though locally it is parallel to adjacent margins. Moreover, the sediments adjacent to the transgressive undersurfaces are disturbed and reconstituted. To explain these features, Francis and Howells (1973) postulated deposition of ash flows in a subaqueous environment, whereby large-scale downsags analogous to load-casts were formed by liquefaction and yielding of the sediments during and after the emplacement of the tuffs. The orientation of the eutaxitic foliation is assumed to be related to the time of collapse relative to cooling.

Eutaxitic structure in welded tuff

The tuffs are cream to pale grey in colour, and are invariably well jointed, though they only locally show good cleavage. They consist mainly of shards, with few crystals, minor lithic debris and matrix. Vitroclastic texture ranges from non-welded through eutaxitic to parataxitic (Beavon and others, 1961).

Capel Curig

South-west of Capel Curig the formation, 120 to 220 m thick, is folded into a broad anticline and comprises three tuff members which are separated by sediments and which are named Garth, Racks and Dyffryn Mymbyr tuffs in upward succession.

The Garth Tuff is characterised by a remarkably discordant base, with apophyses and pipe-like load-casts, up to 100 × 250 m in plan, detached in the underlying sediments. The lower and middle parts of the unit are welded, but there is an upward gradation to evenly bedded, reworked tuff which is concordant with, and locally passes gradually into, the overlying sandstones. Layers of accretionary lapilli are recorded from the bedded upper part of the tuff on the southern limb of the anticline (Figures 4 and 5).

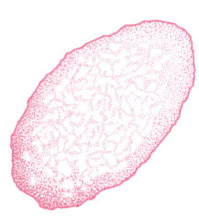
Accretionary lapillus

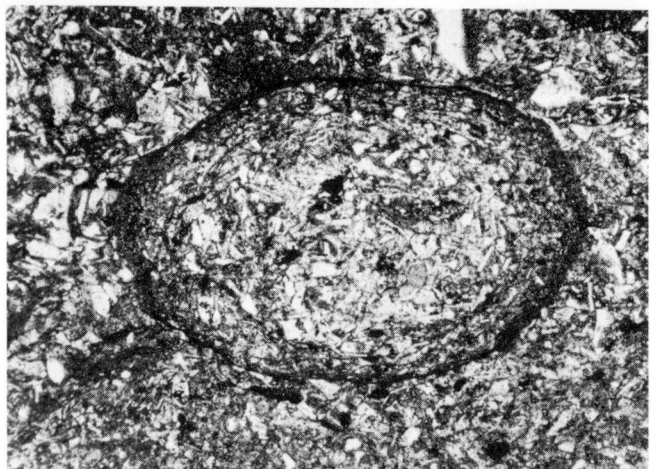

Figure 4
Photomicrograph of accretionary lapillus from the Garth Tuff.
E 35163, ×40

Figure 5
Photomicrograph of welded tuff with eutaxitic fabric. Shards deformed around albite-oligoclase crystal. Garth Tuff.
E 39682, ×40

Figure 6
Photomicrograph of vitroclastic tuff with cuspate shards, bubble walls and fragmented feldspar crystals. Top of the Lledr Valley Tuffs.
E 39162, ×40

The sediments between the Garth and Racks tuffs consist mainly of fine-grained pale green sandstones, up to 60 m thick, which locally yield Soudleyan faunas (see p. 70; also Diggens and Romano, 1968). At the top is an impersistent band of mudstone which is thickest, 18 m, towards the north-eastern closure of the anticline, where it is locally disturbed beneath the irregular base of the Racks Tuff.

The Racks Tuff, up to 50 m thick, varies laterally in lithology from well bedded and unwelded in the west to massive and welded elsewhere. Traced north-eastwards along the north limb of the structure, the tuff, locally welded, becomes discontinuous, first forming isolated pods then vertical and horizontal stringers retaining a welded fabric. Clasts of welded tuff, devitrified glass with perlitic fractures and patches of siliceous nodules are common throughout the outcrop.

Perlitic texture

The Dyffryn Mymbyr Tuff is recognised only on the north-western limb of the anticline where it is separated from the Racks Tuff by 28 m of cleaved grey siltstones and mudstones. It thins north-eastwards and passes from a coarse-grained lithic tuff, south of Cwm Clorad Isaf, to a fine muddy tuffite, north of Plas-y-Brenin. The lower part of the unit is muddy and contains accretionary lapilli. The content of mud and the sedimentary structures suggest that the Dyffryn Mymbyr Tuff is water-laid.

Lledr Valley

The Lledr Valley Tuffs (the local equivalent of the Capel Curig Volcanic Formation) form a faulted outcrop on the flanks of a tight anticline. The lowest unit is impersistent and consists of coarse clastic lithic tuff seen in small outcrops in Gallt Tan-yr-Allt.

The main and most persistent unit, up to 75 m thick, crops out between Lledr Cottage and the River Conway in the steep forest slopes on the north side of Afon Lledr. Correlation along the faulted crop is difficult, for it is apparent that in some places the tuff wedges out and in others its base cuts down into disturbed sediments. The general lithology is closely comparable to that of the Garth Tuff of Capel Curig; the lower and middle parts are massive and welded while the upper part contains undeformed shards, often fragmented, associated with complete bubble forms, clasts of pumice, spherulitic recrystallised tuff and devitrified glass with perlitic fractures. Siliceous nodules, in places occupying the whole fabric, are a distinctive feature. Zones of brecciation, in which angular silicified tuff blocks are associated with patches of siliceous nodules, occur in the gorge of the Afon Lledr, 0.5 km E of Lledr Cottage.

Tuffs of similar lithology to the main unit crop out northeast of Lledr Cottage in three small bodies, oval in plan, transgressing bedded silty sandstones, at a horizon about 30 m above the main unit. In places the bodies have nodular peripheral zones in which welding can be seen to persist almost to the outer margins. The bodies may be the lobe-like remnants of a once continuous ash flow which was partly removed by intraformational erosion. Supporting evidence is provided farther east where an impersistent band of crystal-lithic tuffite crops out at the same horizon.

North and West of Penmachno

South of the Lledr Valley the Capel Curig Volcanic Formation crops out south-west of Ty-Mawr, on the steep slope near Bwlch-y-Maen, and in an outlier north of Iwerddon. Tuffs can be determined at up to three different horizons; because they are faulted and discontinuous they cannot be individually correlated from one area to another.

South-west of Ty-Mawr the massive, white-weathered and strongly welded lowest tuff is folded into a broad anticlinal structure. A similar tuff crops out in the steep valley side, west of Bwlch-y-Maen, and in the shallow synclinal structure north of Iwerddon. About Pigyn Esgob, south-west of Ty-Mawr, at a slightly lower stratigraphical horizon, similar tuff is seen in small oval-shaped outcrops which cross-cut the associated sandstones (Francis and Howells, 1973).

North of Iwerddon two welded tuff units are separated by siltstones and sandstones with an impersistent bed of crystal-lithic tuffite.

Parataxitic structure in welded tuff

South of Dolwyddelan

Discrete bodies of tuff, similar to those described around Pigyn Esgob, and at approximately the same horizon, are the sole representatives of the formation to the south of Dolwyddelan. They occur near the base of the dominantly sandstone sequence that underlies the Snowdon Volcanic Group in this area. Outcrops are irregularly oval in plan and up to 170 m in their longest dimension. Margins are steep and clearly transgress the subjacent sediments. Above Afon Penamnen, nine such bodies have been mapped in a relatively small area. One exposed at Carreg Alltrem [7395 5068] appears to be conformable at the base, where the lowest 2 m of tuff are green, cleaved and crystal-rich, resembling the basal part of the Rolwyd body figured by Francis and Howells (1973, fig. 6). The rest of the tuff is white and massive with prominent columnar jointing and a strong welded fabric (E 38686), generally coincident with the regional bedding.

Nodules occur throughout the smaller bodies and near the margins of the large. The nodules are siliceous, with quartz and chlorite filling septarian fractures. Where observed, the fabric within the nodules is strongly welded parallel to the peripheries of the bodies. Similar nodules occur within the adjacent, often highly disturbed sediments.

Because these bodies resemble the pipe-like bodies of Capel Curig, the Lledr Valley and Rolwyd, it is assumed that they were formed similarly, as detached lobes of a submarine ash flow. As in the Lledr Valley, the parent sheet may have been removed by intraformational erosion and, if so, it is represented now only by a nearly conformable line of lenses of cleaved tuff in Cwm Penamnen.

Strata between the Capel Curig Volcanic Formation and the Crafnant and Snowdon Volcanic groups

Although considerable lateral variation in lithology is apparent in the strata between the Capel Curig and the Crafnant and Snowdon volcanics, they can be subdivided as follows:
3 Siltstones and sandstones, up to 150 m;
2 Sandstones with subordinate beds of acid tuffite, basic tuff and siltstone, up to 225 m;
1 Mudstones and siltstones with subordinate sandstones, up to 600 m.

This subdivision is convenient for purposes of description, though it should be noted that it is difficult to uphold in two areas. One is around Dolwyddelan, where the whole sequence consists of alternating thick siltstones and sandstones in the south, but of little more than siltstones in the north, where only the thin impersistent tuffite beds of 2 serve for correlation. The other is the ground between the Lledr and Llugwy valleys where both tuffites and sandstones in 2 are so thin or impersistent that they are not continuously mappable units.

1 The Capel Curig tuffs are nearly everywhere overlain by sandstone of variable thickness. On the northern flank of the Capel Curig Anticline it is thin and wedges out eastwards: on the southern flank, where the Dyffryn Mymbyr Tuff is absent or is represented only by volcanic detritus, it merges eastwards with the underlying sandstone. In the Lledr Valley up to 120 m of sandstone directly overlies the upper tuff unit east of Pen-aeldroch but it wedges out on the northern side of the closure of the anticline; it becomes more silty when traced eastwards.

The succeeding cleaved, commonly pyritic mudstones and siltstones are thickest (at least 450 m) on the flanks of the Capel Curig Anticline. There, between Cefn-y-Capel and Creigiau'r Gelli to the north, and across the northern and north-western slopes of Moel Siâbod to the south, they form craggy ground on which erosion has distinctively picked out bedding, cleavage and fault lineaments. They have been worked extensively for slates at Rhos Quarry [7285 5635] and, on a smaller scale at two other quarries [7155 5695; 7170 5550]. No macrofossils have been found in this area, nor has treatment of several samples yielded any organic-walled microfossils. To the east, however, a quarry [7837 5501] in siltstone at the side of a forestry road east of Llyn Elsi yielded a predominantly shallow-water molluscan fauna. Other faunas obtained nearby [7865 5596; 7845 5593] are of late Soudleyan to early Longvillian age (p. 70) and resemble the assemblages at the top of the Allt Ddu Mudstones and Lower Gelli-grîn Calcareous Ashes of the Bala District (Bassett and others, 1966). In the southern part of the area, between Penmachno and Dolwyddelan, dominant mudstones give way to a sequence of silty mudstones with up to three thick sandstones.

2 Sandstones with subordinate acid tuffites, basic tuffs and siltstones collectively form a distinctive lithological sequence. In the ground to the north of the district, broadly equivalent strata were termed the Llyn Cowlyd and Bwlch Cowlyd Sandstone formations by Diggens and Romano (1968). The sequence can be traced southwards to Capel Curig, across the Clogwyn-yr-Eryr Syncline and the Pont Cyfyng Anticline to flank the synclines of the southern slopes of Moel Siâbod and of Dolwyddelan.

Because they weather white, the acid tuffites are the most distinctive members of the sequence. They are up to 10 m thick, though commonly 4 m or less, and occupy at least four horizons. At most localities, however, only two – not necessarily the same two – can be seen, so that long-range correlation of individual units is impossible. All are fine-grained rocks, and although a few of the specimens collected approach the composition of tuff, with perfect cuspate shards of about 0.1 mm (E 38722), most are composed of fragmented shards set in a fine matrix which probably represents original volcanic dust with or without a mudstone fraction. The matrix consists either of a fine quartzose aggregate which has made the rock suitable for use as a honestone, or of a recrystallised micaceous aggregate (E 38179) showing cleavage orientation. Small subangular fragments of quartz and feldspar are characteristic of some tuffites (E 35155), and irregular plates of carbonate scattered throughout are a feature of others (E 38176). Slender skeletal rod-like aggregates of ragged chloritic flakes showing no preferred orientation are a persistent microscopic feature throughout the district from north of Capel Curig (E 35156) to Dolwyddelan (E 38179). Cross-bedding is commonly seen, particularly at the tops of units: in places it is disturbed, as at an outcrop [7258 5811] north of the Capel Curig Youth Hostel. The absence of well-defined sorting, the evidence of current-bedding and disturbance at the tops of units and the local impersistence all indicate that the tuffites represent the air-fall of fine ash into shallow water where it was subjected to reworking and to local penecontemporaneous erosion. The almost invariable presence of sandstone above the tuffites further suggests that the sudden arrival of some thickness of ash into shallow water attracted coarse sediment by

Cross-bedding

Figure 7
Sketch of the top of Curig Hill from the south, showing bedded basic tuffs dipping steeply eastwards

shoaling until such time as subsidence re-established the deposition of silt.

Basic tuffs occur at two horizons, the main outcrops being north of Capel Curig. The lower horizon, which lies within the sequence of acid tuffites and sandstones, gives rise to a wide outcrop forming Curig Hill but thins rapidly to north and south (Figures 7 and 8). The prominently cleaved tuffs are chloritised, spilitised and are mainly vitroclastic. They are bedded throughout, although poorly sorted and rarely graded, and are apparently conformable at the top and at the base. Detailed measurements made in the course of 25-in mapping by Dr J. W. Baldock include a semi-quantitative size analysis of lapilli and bombs at 18 localities—all indicating a gradual increase in size towards the centre, near the top of the hill, where there is a marked change to a zone of slumped agglomerate with blocks measuring up to 0.5×1.0 m in diameter over an area some 10 m across. Even more significant are differences which he noted in structure on each side of a plane of discordance which crops out as a median north–south line within the basic tuffs. To the west of it dips are steep and centroclinal, decreasing from $80°$ at the margin to about $50°$ near the coarse slumped central agglomerate; this pattern is modified in the south by a small (penecontemporaneous slump?) fault. East of (that is above) the plane of discordance the dip is uniformly eastward, generally at $25°$ to $30°$, concordant with that of the overlying sediments. This eastern sector includes a lens of fine-grained, aqueously reworked, tuffs and tuffaceous sediments.

Bedded tuffs with marginally steep centroclinal dips are characteristic infillings of the upper levels of funnel-shaped necks or pipes, particularly those which once fed small, short-lived submarine or other maar-type volcanoes (bibliographies and mechanics of subsidence into such pipe-structures are given by Francis, 1970 and Lorenz, 1973). The bedded tuffs which lie above the discordance, passing upwards without a break into younger sediments of the normal succession, are similarly characteristic in marking the final destruction of ash rings and ash cones by post-eruption erosion and sedimentation at the sites of such volcanoes (Francis and others, 1968, pp. 404–405). On this evidence there can be little remaining doubt that Curig Hill marks the site of a former depression infilled by basic vitroclastic material derived from a small underlying volcanic vent.

The upper of the two basic tuffs lies near the top of the uppermost sandstone of the sequence and crops out in the same area north of Capel Curig. It forms a small crag near Bryn Llŷs [7275 5790] and can be traced northwards along the western slopes of Clogwyn-mawr beyond the northern margin of the district. The unit is formed of two components, a lower mudflow and an upper basic tuff, both rather less than 2 m thick, separated by a parting of siltstone. The mudflow consists of angular clasts of sediments and acid tuffs rudely aligned parallel to the local bedding; it is probably the same bed as the lahar mapped discontinuously in the ground around Llyn Cowlyd farther north (Diggens and Romano, 1968). The basic tuff is bedded and contains clasts of basalt and basaltic pumice, both highly chloritised and carbonated, in a fine-grained matrix containing conspicuous amounts of chlorite and ore minerals (E 38373/4). As the unit cannot be traced south of the River Llugwy, a 2- to 3-m basic tuff which is impersistently exposed [7100 5384] at about the same horizon 4.5 km to the south on the slopes of Moel Siâbod, is assumed to represent a deposit from a different volcanic centre.

The sandstones of the sequence are generally greyish green and well bedded, and are locally rich in shelly faunas. The bed overlying the lowest acid tuffite on the north side of Curig Hill is rich in *Plaesiomys multifida* Salter (= *Dinorthis multiplicata* Bancroft) and is possibly the equivalent of the *Multiplicata* Sandstone mapped farther north around Llyn Cowlyd (Diggens and Romano, 1968), though there is evidence to show that the fossil is not restricted to one horizon. Towards the top of the sandstone sequence, north of Coed Bryn-y-Gefeiliau, a calcareous sandstone [7411 5691] yielded a fauna of probable Longvillian age. The Soudleyan–Longvillian boundary thus lies within the sequence — probably nearer the base than the top on the evidence of forms obtained from Dolwyddelan (p. 70; Romano and Diggens, 1969).

In the ground south-west of Betws-y-Coed the sequence is ill defined. Three sandstones are recognised north-east of Mynydd Cribau, but they pass laterally into a sequence dominated by siltstones and slates farther east. Slates underlying the lowest local sandstone have been quarried at Hafod-lâs, near Betws-y-Coed, where they include a tuffaceous band, less than 30 cm thick, composed of talcose white mica and highly altered crystals of feldspar (E 36235). On the west side of Mynydd Cribau the highest sandstone is underlain by a white-weathering band of tuffite (E 37201) which does not appear to persist along the strike.

Figure 8
Map and section of the basic tuffs at Curig Hill

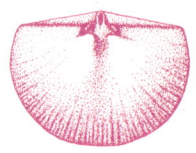

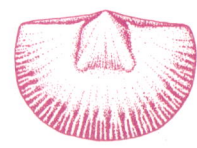

Brachiopod
Plaesiomys multifida
½ life size

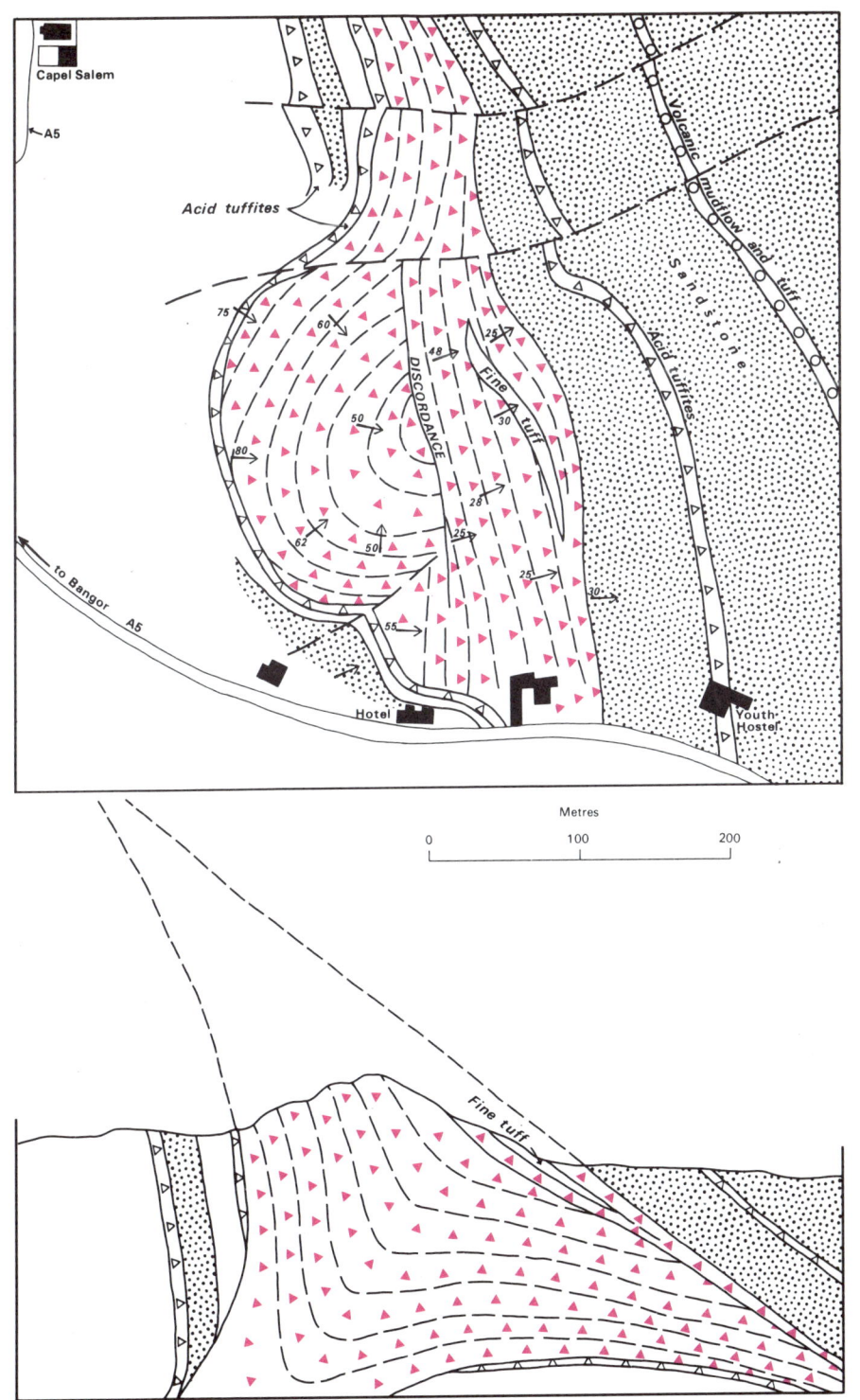

3   To the north of the district, around Llyn Cowlyd, Diggens and Romano (1968) divided the argillaceous strata at the top of the 'Glanrafon Beds' into a lower Pen Llithrig-y-wrâch Siltstone Formation and an upper Marian Mawr Mudstone Formation. They state that the Crafnant Volcanic Group unconformably overlies these beds, progressively cutting out the mudstone so as to rest on the lowest 45 m of siltstone at the northern margin of the present district. This unconformity has not been confirmed by the recent survey, which has shown both divisions (not shown separately on the published map) to persist, without reduction in thickness of the upper mudstone division, throughout the ground around Capel Curig. Moreover, although these argillaceous beds are much thinner in the southern part of the district, a siltstone/mudstone boundary can also be traced around the Dolwyddelan Syncline, where the mudstones have been worked for slate in the Pen Llyn and Rhiw Goch quarries.

The twofold division is less apparent around Betws-y-Coed, where a thick wedge of rhyolite crops out low on the slopes of Coed Aber-llyn, west of Hafanedd. Although the rhyolite is highly silicified, with the original fabric largely obliterated by a fine quartzose mosaic, flow-banding can still be distinguished, particularly in the northern exposures. To the south the rock is highly brecciated and the angular fragments of rhyolite show distinctive spherulitic recrystallisation of quartz and chlorite controlled by the flow-banding (E 35629).

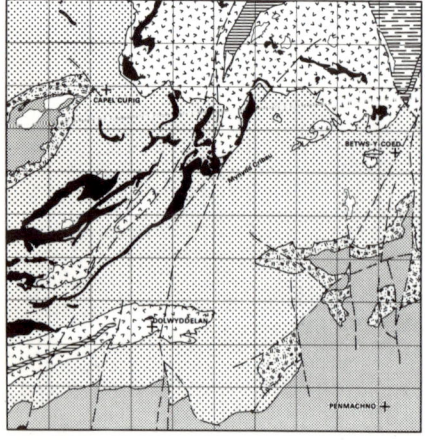
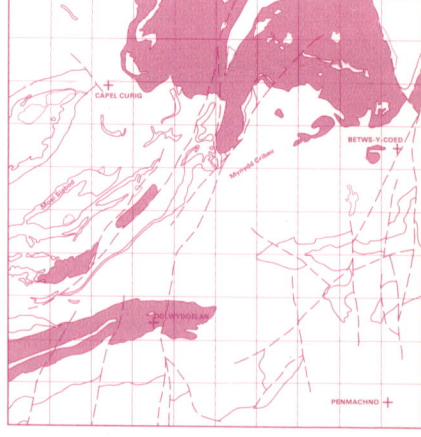

Crafnant and Snowdon Volcanic groups

3

Late Caradoc volcanism produced large volumes of eruptive rocks which originally covered the whole district. Complete sequences are now found, however, only in the Dolwyddelan Syncline and in the area north of the Afon Llugwy, between Capel Curig and Betws-y-Coed, though in the ground between those areas lower members of the sequence are preserved in small synclinal outliers (Figure 9).

The Dolwyddelan rocks were correlated by Williams and Bulman (1931) with the succession in Central Snowdonia (Williams, 1927) and named the Snowdon Volcanic 'Suite' (now Group), comprising three divisions, namely Lower Rhyolite-Tuffs, Bedded Pyroclastic 'Series' and Upper Rhyolite-Tuffs. The rocks north of the Llugwy were similarly correlated with the Snowdon volcanics by Davies (1936), who applied the local term Crafnant Volcanic 'Series'. Both nomenclatures are retained in this account, modified only to meet modern requirements of stratigraphical terminology. In addition the Crafnant Volcanic Group has been divided into Lower, Middle and Upper formations (Howells and others, 1971). The lower formation has been correlated in some detail with the lower formation of Dolwyddelan (Figure 18; Howells and others, 1973). The middle and upper formations of the two areas can also be broadly equated, though they show more lateral variation in lithology than the lower formation. In particular, the essentially basic nature of the Bedded Pyroclastic Formation in Central Snowdonia becomes markedly less pronounced when traced from west to east within the Dolwyddelan Syncline, while the equivalent Middle Crafnant Volcanic Formation is almost entirely acid in composition.

Soudleyan faunas have been recorded at or near the base of the group in Central Snowdonia (Harper *in* Shackleton, 1959; Williams and Harper *in* Beavon, 1963), whereas in the Capel Curig and Betws-y-Coed district the Soudleyan–Longvillian boundary is firmly drawn within the sandstone-tuffite assemblage some distance below the base of the group (p. 3). It is uncertain whether this implies an earlier start of the main volcanism in Central Snowdonia or anomalies in the preservation and collection of specimens. The Crafnant–Snowdon Volcanic Group yielded few fossils during the recent survey, but its mid-Caradoc age is established by the *Diplograptus multidens* Zone fauna contained in the Middle Crafnant Vol-

Extent of the Crafnant and Snowdon Volcanic groups

canic Formation (p. 70), and in the black slates overlying the Group at Dolwyddelan and Betws-y-Coed.

Black slates occur below the uppermost formation of both the Snowdon and Crafnant Volcanic groups.

In the following account the two areas are described separately under local formational headings.

Figure 9
View westwards up the Llugwy Valley with Moel Siâbod in the distance

Crafnant Volcanic Group

The Crafnant Volcanic Group crops out in the high ground south of the Afon Llugwy, between Clogwyn-mawr in the west and Llyn-y-Parc in the east. Its subdivision into Lower, Middle and Upper formations is based on distinctive volcanic lithologies. The Lower Crafnant Volcanic Formation, up to 210 m thick, is composed of three units of ash-flow tuffs, up to 60 m thick, separated by slates, siltstones, a few thin limestones and local hyaloclastites and basic tuffs (Howells and others, 1973). The Middle Crafnant Volcanic Formation, 30 to 120 m thick, comprises alternating thin acid tuffs, tuffites, siltstones and mudstones which in places are highly admixed and disturbed. The Upper Crafnant Volcanic Formation, 60 to 150 m thick, is an ill-sorted pyroclastic deposit which is characterised by much included argillaceous debris.

Lower Crafnant Volcanic Formation

The following account summarises the detailed description of the formation already published by Howells and others (1973). The tuffs of the component units, numbered 1 to 3 from the base upwards, form scarp features (Figure 10) which extend westwards from Clogwyn Cyrau, in the forest near Betws-y-Coed, to Capel Curig, where they strike northwards along the ridges of Clogwyn-mawr, Clogwyn Cigfran and Crimpiau. They are predominantly massive, poorly cleaved grey rocks, weathering white. At some localities the basal rocks of units are mica-rich, bluish grey and well cleaved. The tops of Nos. 1 and 2 units are uniformly flinty.

In thin section these rocks are seen to be acid vitric tuffs with a variable content of crystals and lithic clasts. Shards are completely devitrified and their shape, independent of size, varies from broad and massive to delicate and spiky: they are commonly tectonically deformed, but are not welded. Crystals, consisting dominantly of sodic plagioclase, with subordinate amounts of fragmented euhedral quartz, are found only in units 1 and 2. Some of the plagioclase crystals are fragmented, others are marginally resorbed. Lithic clasts are common in all units and include fossil fragments, siltstones,

Fractured feldspar crystal

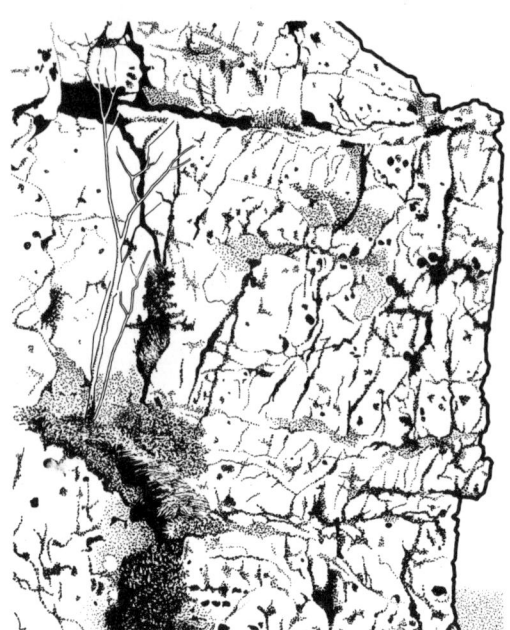

Figure 10
Sketch of crag of massive tuff in the Lower Crafnant Volcanic Formation

acid tuff, pumice, basic tuff and dolerite. The matrix is a fine-grained admixture of quartz, feldspar, sericite and chlorite probably representing an original fine vitric dust. Carbonate concretions are randomly distributed throughout all three units, but siliceous concretions are peculiar to the top of No. 2 Unit.

No. 1 UNIT   This unit varies in thickness from 21 m at Aber-llyn [7941 5730] to 56 m at Capel Curig [7270 5760]. The regional setting suggests conformity with the underlying sediments, though the base is exposed only at Cae-mawr [7567 5716] where the tuff locally cuts down into underlying siltstone to a depth of 1 m.

Through most sections west of Cae-mawr a general upward fining is apparent and subdivision can be made into a lower sub-unit with crystals and sedimentary clasts, a central sub-unit with small pumice clasts and few crystals, and an upper sub-unit of finer debris. Bedding is generally a distinctive feature of the unit away from the area of Allt Goch [7443 5760] and Capel Curig. To the north of Capel Curig, thick regular beds, separated by silty bands with well-developed cleavage, occur from 7 to 21 m above the base.

Lithic clasts are generally prominent in the lower sub-unit, especially at Pen-yr-allt-uchaf [7864 5728], where blocks up to 1.3 m across occur. Carbonate concretions up to 0.7 m are common and in places, as at Cae-mawr and Towers [7578 5758], show a tendency to be concentrated near the top of the lowest sub-unit and also to coalesce.

Strata between Nos. 1 and 2 Units   These strata generally consist of siltstone and mudstone less than 30 m thick although up to 70 m of strata, including a thick mass of hyaloclastite, are present north of Crimpiau.

Exposures of the sediments are mainly small and isolated. They include, near the base of the sequence, an argillaceous limestone rather less than 1 m thick, seen in a forestry road section [7445 5788] at Allt Goch, to the south of Cae-mawr [7526 5666] and in the small synclinal outlier in Coed Maes-newyddion [7740 5691]. Longvillian faunas have been determined from the first two of these exposures.

The well-exposed hyaloclastite north of Crimpiau includes blocks of albitised and chloritised basalt (E 38376) in a matrix of carbonate and chlorite, with fragments of vesicular basaltic glass, replaced by chlorite and ilmenite with sphene still discernible (E 38377).

No. 2 Unit   The thickness of No. 2 unit ranges from 18 m at Cae-mawr to 73 m at Pencraig [7662 5784]. At Pencraig a sharp contact with the underlying mudstones is exposed.

In the west the unit is generally uniform and massive, whereas eastwards bedding and ribbing are more apparent and pumice clasts, up to 20 cm across, occur. In the lowest 3 m of the unit brachiopod casts are common at Cae-gwyn [7599 5820] and ooliths at Aber-llyn.

Crystals of sodic feldspar and quartz are present throughout the unit, but are concentrated in the basal part in most sections. Siliceous concretions are common and at the western extremity of outcrop the unit is characterised by an overall fine flinty siliceous character.

Figure 11
Sketch of typical Lower and Middle Crafnant scenery around Llyn Bodgynydd

Strata between Nos. 2 and 3 Units   These strata consist of slates and siltstones rarely exceeding 30 m, but not completely exposed at any locality. Above the scarp on the north side of Afon Llugwy, near Pen-yr-allt-uchaf, small exposures show the lower beds to be distinctly tuffaceous, with rich Longvillian faunas. These beds are overlain by brittle blue cleaved slates up to the base of No. 3 Unit.

Farther west, near Waenhir [7370 5860], a basic tuff with layers of mudstone pellets, up to 2 cm in diameter, can be traced in small crags below the No. 3 Unit. Similar thin layers of basic tuffs are interbedded with convoluted siltstones below the thick dolerite sill on Creigiau Geuallt [7302 5894].

No. 3 Unit This unit ranges in thickness from 40 m on the ridge north-north-east of Capel Curig to approximately 20 m at Cae-gwyn. It is generally heterogeneous, being composed of vitric tuff, often rich in clasts, with tuffites and silty intercalations.

At Capel Curig bedding is seen only 21 m above the base of the unit in a thin agglomeratic band composed of rounded clasts with interstitial lithic material. Northwards and eastwards from Capel Curig thin discontinuous crenulated ribs, accentuated by chlorite smears parallel to the bedding, are a feature of the unit. They are best displayed on the weathered lower surfaces at Allt Goch.

Except in the finer grained layers, lithic clasts are common and include andesite, hyaloclastite, basic tuff, vitric acid tuff, pumice and siltstone. The unit is also distinguished from Nos. 1 and 2 units by the absence of xenocrysts and the remarkable admixture and range of shard sizes.

Interpretation The interpretation of the emplacement of the Lower Crafnant Volcanic Formation is given at the end of the section dealing with the equivalent Lower Rhyolitic Tuff Formation (p. 36).

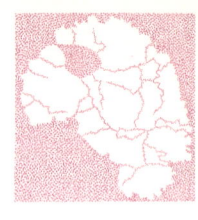

Quartz xenocryst

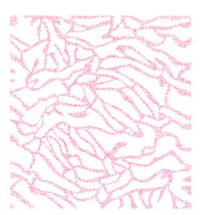

Vitroclastic texture in tuff

Middle Crafnant Volcanic Formation

The Middle Crafnant Volcanic Formation comprises alternations of tuffs, tuffites and sediments. The range of lithologies is well seen in the type area, between Llyn Bodgynydd and Coed-mawr [7854 5840], where the formation, characteristically even bedded and flaggy (Figures 11 and 12), includes bluish black mudstones, mudstones with scattered idiomorphic feldspar crystals, tuffites, muddy tuffs and pale grey vitroclastic tuffs with lithic blocks.

SEDIMENTS   The sediments consist predominantly of mudstone and siltstone with infrequent narrow ribs of coarse sandstone. The siltstones (E 35197) contain varying amounts of iron ore and a few feldspar fragments, up to 0.5 mm, in a fine matrix of sericite, chlorite and an aggregate of quartz with some feldspar. Occasionally the siltstones contain a distinct chlorite fraction (E 37180), both finely disseminated and re-crystallised in vermicular growth.

Sole markings characteristic of turbidites are recognised at the bases of the few sandstone layers. The sandstones fill scour structures in the underlying mudstones and these show evidence of subsequent deformation with, in places, lobes of sandstone almost completely isolated by the 'flames' of the underlying mudstone. The sandstone shows normal grading with the coarsest fraction and a distinctive concentration of pyrite in the scoured hollows (Sanders, 1965, after Natland and Kuenen, 1951) (Figure 13.1).

Figure 12 Forestry road section in tuffs, tuffites and black mudstones showing even bedding and low dips typical of the Middle Crafnant Volcanic Formation in the Sarnau area

TUFFITES   The tuffites are typically fine-grained, banded rocks that differ macroscopically from the associated mudstones in their pale grey to white colour, which is particularly apparent on weathered surfaces. Gradations can be seen from dark grey mudstones to fine-grained, pale grey tuffites. The finer tuffites vary in grade from chert-like to fine sandstone. Small pale patches which commonly break the homogeneity are composed of well-formed cuspate shards. Matrices consist of fine quartzo-feldspathic aggregates with sericite shreds and finely disseminated chlorite. The dark grey bands, which give the rocks a striped character, result from concentrations of chlorite, an opaque carbonaceous fraction and iron ore.

Coarser tuffites, which generally grade into tuffs, consist of small rounded albite crystals, up to 1.2 mm, in a dark grey base consisting of quartz, feldspar, shreds of chlorite, ragged fragments of carbonaceous material and iron ore. Shard-rich areas indicate the close admixture of epiclastic and pyroclastic material.

Post-depositional structures are a characteristic feature of the tuffite sequences. Alongside a forestry road [7698 5926], near Bryn-y-fawnog, a massive tuff overlies a tuffite which is

Figure 13
Depositional and post-depositional structures in the Middle Crafnant Volcanic Formation, Sarnau area
1 Turbidite sandstone band with pyrite-rich basal layer infilling scour structures in underlying black mudstone. Scour structures show evidence of deformation on right-hand margin of specimen.
2 Bedded tuff, tuffite and mudstone showing thixotropic yielding in central band with adjacent bands undeformed.
3 Bedded tuff, tuffite and mudstone showing cross-lamination in lower unit and extreme deformation of the base of a fine tuff layer with upward injection of mudstone near the top.

severely deformed at the base by upward flame-like incursions of the underlying siltstones (Figure 14.1, 2); fine, even bedding in the siltstone is contorted into the flames. Here the upward transgressions of the siltstone are not all sharply pointed; broad upwarping also occurs.

Convolute laminations in the tuffites can be examined in the scarp features [7755 5884] south of Sarnau (Figures 13 and 14.4). The contortions are confined to the cherty tuffitic layers within evenly banded tuffs and tuffites. In some instances [7814 5848] the convolution is associated with small penecontemporaneous faults which clearly dislocate the evenly bedded layers below, but cannot be traced upwards through the zone of convolution.

TUFFS These rocks range from tuffs to tuff-breccias in grade (Fisher, 1966) and from well sorted to ill sorted. Perfect cuspate shards in a good vitroclastic fabric (E 35192) are typical, and included fragments of crinoid columnals and a graptolite fragment have been recorded.

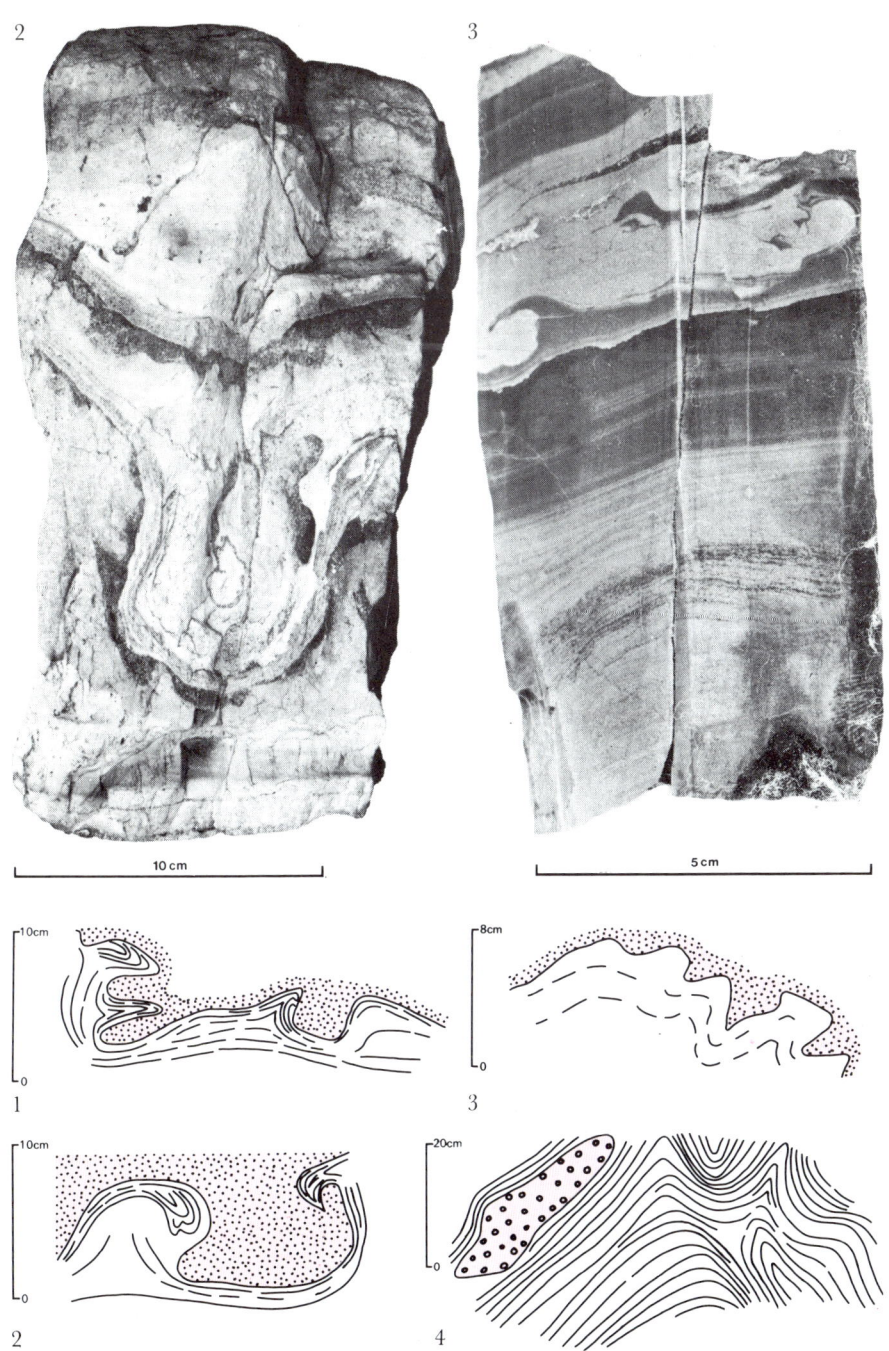

Figure 14
Field sketches of post-depositional structures in the tuffites of the Middle Crafnant Volcanic Formation

The coarsest tuffs form massive beds, up to 1.3 m thick, in distinctive scarp features south-west of Bryn-y-fawnog. They contain isolated blocks, up to 0.5 m in diameter, of dark grey indurated mudstone which show no preferred orientation or position within the beds. Small vesicle-like cavities are distinctive on the exposed surfaces and are formed by the weathering of ill-defined plates of carbonate. A good vitroclastic texture (E 37181) is slightly obscured by a fine recrystallised quartzo-feldspathic mosaic. Blocks of mudstone and tuffite, up to 0.2 m in diameter, occurring in coarse tuffs west-south-west of Sarnau, show faint internal planar structures which closely reflect an indented periphery and indicate that the blocks were unlithified at the time of incorporation (Figure 14.3).

As in the sediments and tuffites, loading structures are a common feature of the tuffs. A specimen (Figure 13) collected from a roadside exposure [7814 5848] shows three units—a lower finely laminated tuff, a central tuffite with planar laminations and an upper banded tuff-tuffite unit. The base of the upper unit is grossly distorted by loading and the main downwarping lobes are almost detached by the upward intruding flames of tuffite.

The evenly bedded, alternating lithologies of the type area can be traced in a steeply dipping succession through the forests to a fault striking south-south-west from Llyn Bodgynydd. North-east of Capel Curig the lowest beds of the formation are well exposed across a broad anticlinal structure west of Afon Abrach. Here, near the base of the formation is a thinly bedded, basic tuff exposed intermittently as far west as the margin of the thick dolerite sill south-east of Creigiau Geuallt.

Higher in the formation, in this area north-east of Capel Curig, the alternations of tuff, tuffite and siltstone are more difficult to distinguish. This is not entirely a function of exposure. On the heather-covered slopes south-west of Clogwyn Manod and on Ffrith Newydd, for instance, the numerous small exposures show such a lack of lateral continuity in lithology as to indicate some breakdown in the bedding. Similarly, in exposures on the dip-slopes about Nant Geuallt, south-west of Clogwyn Manod, there is a broad and randomly distributed range of tuff types from coarse agglomerates (E 38868) with rounded clasts of spilitic basalt, altered basaltic glass, perlitic fractured rhyolites and chloritised shale, to fine, cream-coloured, cleaved vitroclastic tuffs. Siltstone intercalations, which are distinctly tuffaceous, are also generally impersistent.

The highest beds of the formation in the Sarnau area are bluish black graptolitic slates (p. 70) which crop out around the lake and in the road cutting north of Sarnau Cottage. They can be traced below the scarp features of the Upper Crafnant Volcanic Formation in Coed Bwlch-yr-Haiarn and Craig y Fuches-lâs overlooking the Nant Gwydir valley.

Perlitic texture in rhyolite

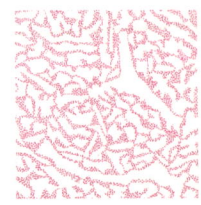

Vitroclastic texture

Pumice fragment

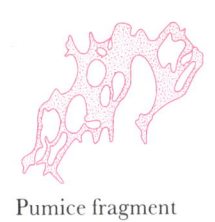

Interpretation The Middle Crafnant Volcanic Formation accumulated in deeper water than did the lower and middle parts of the Carneddau Group. There is a predominance of mudstones and fine siltstones in the sequence and the few sandstones that occur show the characters of turbidites transported from a distant source.

The characters of the volcanic rocks are quite different from those of the Bedded Pyroclastic Formation of Dolwyddelan and Central Snowdonia, suggesting eruptions from different centres. The pyroclastic material of the Middle Crafnant Volcanic Formation probably emanated from a centre to the north. It may represent continued activity at the source of the highest ash flow of the Lower Crafnant Volcanic Formation (Howells and others, 1973). The transition from well-bedded to slumped sequences as the formation is traced from the Sarnau area to the ground north-east of Capel Curig may reflect passage across the margin of an unstable volcanic pile.

Around Sarnau the coarse tuffs are very different in grade from the associated sediments. The blocks of originally unlithified mudstone and tuffite, the high clay content in parts of the vitroclastic fabric, and the marine fossils indicate emplacement within a marine environment. The tuffs presumably originated in shallower parts of this environment as small ash flows or secondary slurries of unstable pyroclastics, and flowed into the deeper parts, incorporating sediment during transport.

The tuffites are essentially fine-grained, pale grey and evenly banded. The banding suggests the settling of fine material through a water column to a level below the wave-base. Where these deposits are thin and directly overlie coarser tuffs, it is reasonable to assume that they accumulated from dust clouds released into water during the transport and emplacement of the tuffs. The thicker deposits are possibly accumulations of fine shardic dust-rich eruptions which settled through the water after air transport from the source area.

The convolute bedding in narrow bands can be explained by the rapid emplacement of tuff on water-saturated, partly lithified tuffite which would have yielded thixotropically. The mechanism producing the convolution in thicker bands is not always apparent. As the internal folds clearly occur in beds that are not themselves folded it is suggested that the load-failure mechanism could have been initiated by earthquake shocks characteristic of volcanic environments.

Upper Crafnant Volcanic Formation

This formation is equivalent to the Upper Tuff Bed of Davies (1936) in the Crafnant country, to the north of Capel Curig. In the district described here it is well exposed in the high forest ground west of the Llyn-y-Parc Fault and across Mynydd Bwlch-yr-Haiarn, north of Sarnau. It can be traced into the

Figure 15 Photomicrograph of muddy crystal tuff with squat cuspate shards and oligoclase crystals set in a matrix of sericite, chlorite and irregular smears of iron ore. From the Upper Crafnant Volcanic Formation. E 35261, ×45

fault-bounded synclinal structure around Llyn Goddionduon and on to the ridge of Clogwyn Manod and Ffrith Uchaf, north-east of Capel Curig where, however, it is difficult to delimit from the Middle Crafnant Volcanic Formation. The upper part of the formation is exposed only in the syncline about Llyn Goddionduon.

Typically the formation is massive and almost entirely without internal bedding structures, so that it forms only crude scarp features. It consists of tuffite composed of a heterogeneous admixture of pyroclastic and epiclastic material. The pyroclastic material consists of cuspate shards of highly variable form, from coarse bubble walls to small fragmented rods and spikes, together with fragmented, rounded and resorbed crystals of sodic feldspar and pumice fragments and blocks. The predominant epiclastic material is muddy in character and consists of a fine aggregate of chlorite, sericite, dusky carbonaceous material and iron ore. Lithic clasts are generally patchy in distribution, as north-east of Llyn Bychan [7537 5958] and include siltstone, acid tuffs, dolerite and spilitic basalt (E 35261) (see Figure 15).

The proportions of the various constituents is highly variable. In places the rocks are tuffaceous mudstones with a high proportion of the muddy fraction and few small fragmented shards and crystals, as seen (E 35217) low in the formation south of Llyn Goddionduon. At about the same horizon, on the north side of the dam at Llyn Bodgynydd, a disturbed sequence of tuffaceous mudstone, tuffite and muddy tuff with blocks of vitric tuff is exposed. In the tuffite (E 38918) the bedding is clearly distorted and the high proportion of mudstone in the sequence gives rise to a strong cleavage. Impersistent ribs of mudstone within a heterogeneous admixture (E 35195) give crude indications of bedding. Smears of fine mudstone are often accentuated by a concentration of iron ore as are many of the cross-cutting cleavage planes (E 37163). Clearer indications of bedding are entirely absent. Carbonate nodules have been recorded in a few localities. In a forestry road cutting near the southern end of Llyn Bychan [7517 5921] a strongly cleaved muddy crystal tuffite contains carbonate nodules, up to 0.7 m in longest dimension, which overprint and slightly deflect the cleavage planes. These nodules are almost entirely decomposed into a soft friable ferruginous mush, the hard cores of which are seen in thin section (E 35216) to consist almost entirely of carbonate obscuring the original fabric.

Interpretation The pyroclastic component of the lithological admixtures in the formation consists of coarse shards and crystals which are typical products of ash flows. The complete lack of sorting and minimal internal bedding features suggest deposition by a high-density turbid flow produced by the remobilisation of pyroclastic material that had been previously rapidly emplaced on unlithified muds.

Snowdon Volcanic Group

The Snowdon Volcanic Group crops out in the Dolwyddelan Syncline and in two synclinal outliers on the southern flank of Moel Siâbod. The sequence is complete in the Dolwyddelan Syncline where the group, 270 m thick, is overlain by black slates (Williams and Bulman, 1931), but in the outliers on Moel Siâbod only incomplete sections of the Lower Rhyolitic Tuff Formation are preserved (Howells and others, 1973).

Lower Rhyolitic Tuff Formation

Three units, termed A, B and C from the base upward are recognised within the formation at Dolwyddelan, though because of lateral impersistence the only locality at which all three are represented is Tan-y-Castell, on the northern limb of the syncline. Only two units (A and B) are present at Blaenau Dolwyddelan and two (B and C) at Adwyr Dwr, while on the southern limb there is only one (B). Two tuff units (B and part of C) occur in the Moel Siâbod outliers. Units B and C are correlated respectively with Nos. 1 and 2 units of the Lower Crafnant Volcanic Formation. Unit A of Dolwyddelan does not extend into the Crafnant country, nor is No. 3 Unit of Crafnant represented at Dolwyddelan (Figure 18).

UNIT A At Dolwyddelan this unit thins from 43 m at Blaenau Dolwyddelan to 10 m at Tan-y-Castell (Figure 16). Farther east it passes into grey slates with lenses of tuff before dying out. Characteristically well bedded, it contains lithic clasts, which are prominent in the west and considerably less common to the east. Sodic feldspar, in clusters and individual

Figure 16 Nearly vertical top surface of Unit A, Lower Rhyolitic Tuff Formation, Dolwyddelan, showing pseudo-cleavage mullions produced by erosion and weathering along cleaved zones

crystals, forms a high proportion of the rock. Delicate shards occur in a matrix of fine-grained chlorite, sericite and quartz. At Blaenau Dolwyddelan the top of the unit is composed of recrystallised acidic clasts.

At Moel Siâbod the unit varies in thickness from 39 m in the western outlier to 29 m in the eastern. It conformably overlies siltstones and fine sandstones and is generally well bedded. In the west it comprises a lower well-bedded crystal-rich vitric tuff overlain by flaggy-bedded silty tuffites which are graded and pass upwards by progressive enrichment in crystals into crystal tuffs. They contain a few randomly distributed rounded rhyolitic clasts and form the top few metres of the unit. In the east the central part of the unit is composed of tuffaceous siltstones, and some of the crystal-rich tuffs show crude grading and diffuse lamination.

Siltstone clasts are common within the tuffs. Crystals, up to 8 mm in length (E 39299), are predominantly of sodic plagioclase; many have a distinctive interrupted twinning and they are commonly sieved with chlorite. The siltstones and plagioclase crystals are distinctive features of correlative value in establishing that the unit is restricted to the south-western part of the district (Figure 18).

Figure 17 Sketch of Moel Siâbod from Dolwyddelan: the syncline formed of Lower Rhyolitic Tuff Formation forms the slopes in the left middle distance

STRATA BETWEEN UNITS A AND B In the Dolwyddelan Syncline these two units are separated by cleaved siltstone and mudstone ranging in thickness from 7 m at Blaenau Dolwyddelan to 16 m at Tan-y-Castell. On Moel Siâbod the strata are up to 10 m thick and are well exposed in the western outlier. They consist of siltstone, fossiliferous and tuffaceous in parts, with a 1-m band of highly altered hyaloclastite (E 39309).

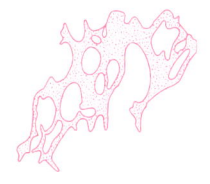

Unit B   At Dolwyddelan this unit varies from 30 m at Minffordd to approximately 65 m at Blaenau Dolwyddelan. Where seen, as at Adwyr Dwr, the contact with the underlying mudstone is sharp, and flames of mudstone, up to 9 m high, intrude the base of the tuff.

In places the basal 2.5 m are evenly banded. Even, thick flaggy bedding is characteristic throughout the unit although parts are massive. Clasts of mudstone and sandstone are numerous at the base and angular pumice fragments are common throughout. The unit is typically crystal-rich with common euhedra of sodic feldspar, often fragmented, and less common subrounded quartz crystals.

Shards show a wide range of size and shape. At the base of the unit, where the matrix is dominantly chlorite and sericite, the recrystallisation of the shard material is often micaceous. At higher levels both shards and matrix are generally more siliceous.

In both outliers on Moel Siâbod the unit is incompletely preserved. In the west the base is irregular and is locally formed of agglomerate. The unit is composed of well-bedded vitric tuff which becomes progressively more muddy towards the top. The lower parts are generally massive with evidence of crude upward grading and faint cross-bedded silty laminations. The upper parts consist of distinctive, flaggy alternations of cross-bedded vitric tuffites with channel features, and thin evenly bedded air-fall vitric tuffs.

The tuffs are typically shard-rich with minor proportions of sodic feldspar crystals and lithic clasts. In the east, crystals and mudstone clasts are concentrated at the base of the unit in association with carbonate concretions, up to 25 cm long.

Pumice fragment

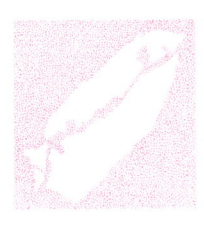

Fractured feldspar crystal

Strata between units B and C   Apart from a section at Adwyr Dwr, these strata are poorly exposed. They consist of pale grey siltstone in beds up to 2 m thick, with bands of crystals and pumice near the base. Locally, they are richly fossiliferous.

Unit C   This unit thins westwards from 24 m at Adwyr Dwr to 15 m at Tan-y-Castell. At the latter locality the inverted base shows shallow channel casts. Bedding varies from poor to well defined with the fining of grade. Dispersed ooliths have been recorded from the basal parts of the unit. Lithic clasts are generally scarce.

Shards vary in size, shape, recrystallisation and concentration. At the base at Adwyr Dwr the fabric has been strongly overprinted by the recrystallisation of chlorite and sericite, and here the shards show strong mutual interference. At the top of the Tan-y-Castell section the rock is finely banded and composed of small fragmented shards with worm casts showing internal structure.

Interpretation   On the basis of established correlation (Howells and others, 1973) the Lower Rhyolitic Tuff Formation and Lower Crafnant Volcanic Formation are here treated together. The combined formation comprises four units of which Unit A is restricted to Dolwyddelan and Moel Siâbod, units B and C of Dolwyddelan and the respective equivalents Nos. 1 and 2 units of Crafnant occur throughout the district, and No. 3 Unit is restricted to the north-east (Figure 18).

The distribution and thickness variation of the lowest unit indicates a westerly source. The poor sorting at Blaenau Dolwyddelan implies deposition from a series of slurries, whereas the reversed grading at Moel Siâbod suggests deposition from suspension flows. The concentration of pebbles at the top of the unit at Moel Siâbod indicates derivation from a local shallow-water environment.

Units B and C at Dolwyddelan and their correlatives, Nos. 1 and 2 units of the Lower Crafnant Volcanic Formation, show little thickness variation across the district. This, together with their broad lateral extent, ill-defined sorting and overall upward grading are characters of flow and rapid emplacement. The lack of induration of included clasts, the preservation of fossil fragments and the absence of welding, further indicate that the flows lost most of their initial heat during transport. The bedding, which suggests water suspension, probably results from a loss of energy within the main flow, giving rise to the formation of sub-flows.

Fiske and Matsuda (1964) described accumulations of unwelded material in a subaqueous environment in the South Fossa Magna, Japan, and ascribed them mainly to slurries and turbidity currents originating from unstable accumulations on the flanks of a submarine volcano. In contrast Unit B/No. 1 and Unit C/No. 2 are single pyroclastic flows with no indication over the area of a composite nature. Because such massive units are not likely to have accrued from the flowage of earlier accumulated material, it is necessary to postulate erup-

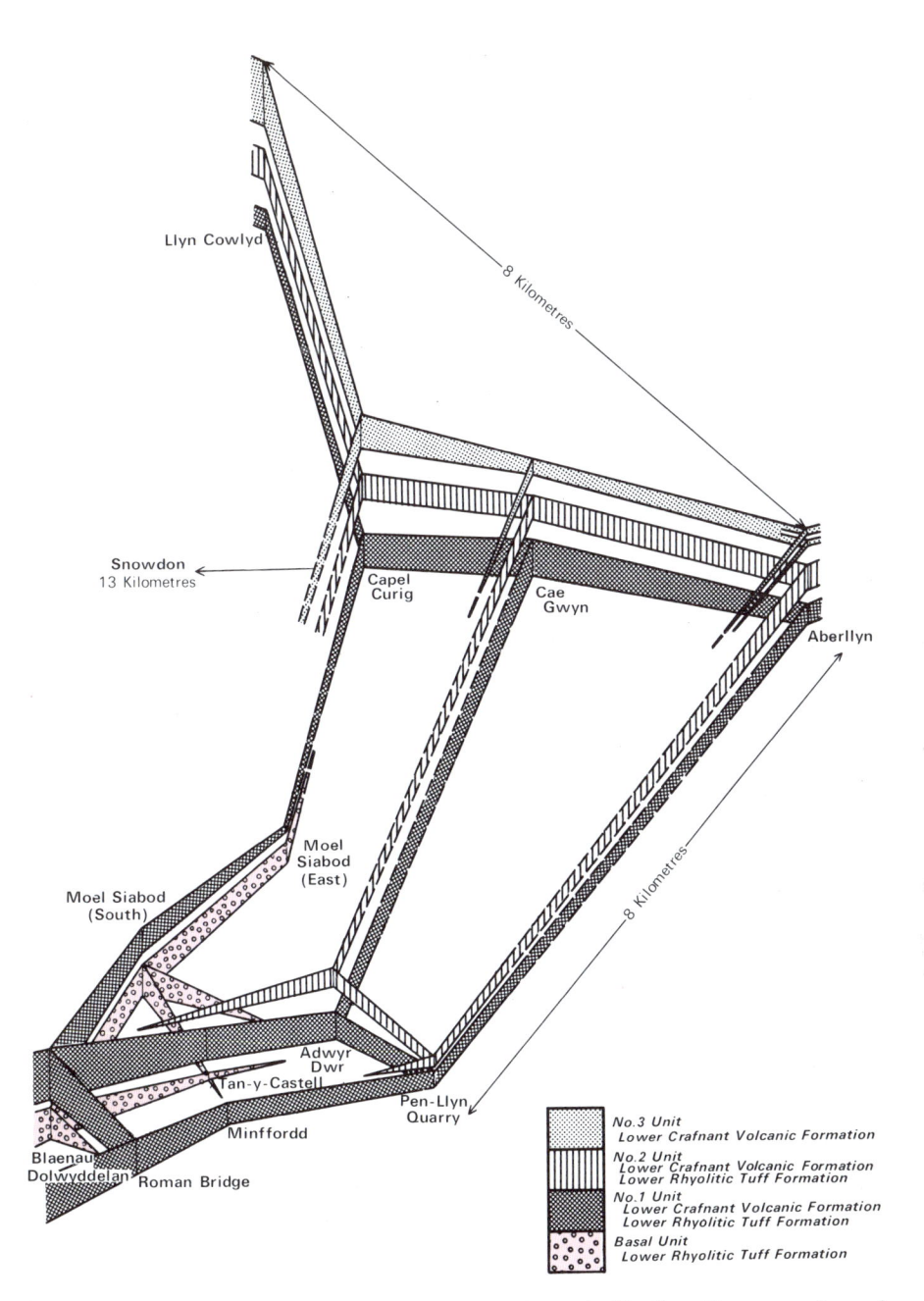

Figure 18 Correlation of the Lower Crafnant Volcanic and Lower Rhyolitic Tuff Formations

tion as the source of the flow. At Penllyn Quarry a cleaved banded micaceous basal zone is interpreted as a base surge deposit which has been cut into by the overlying massive tuff. At Moel Siâbod thin beds of graded laminated air-fall tuffs associated with reworked volcanoclastic deposits indicate an emergence of eruptive source and a shallowing of the sea.

The source area for these two widespread flows has been postulated to lie in Central Snowdonia (Howells and others, 1973). Of the continuous volcanic emissions in that area they are the only major flows which seem to have escaped eastwards.

The characters of the No. 3 Unit are more suggestive of composite flows emplaced as slurries, fluxoturbidites or the proximal parts of turbid flows. The unit differs from the underlying flows in lacking crystals. Moreover, it does not extend south of Afon Llugwy. It thus has a different source—one which probably lay fairly near and to the north.

Bedded Pyroclastic Formation

This formation is best exposed on the northern limb of the Dolwyddelan Syncline, where its thickness varies from 180 m at Blaenau Dolwyddelan to 60 m in the vicinity of the castle. It consists of well-bedded calcareous tuffs, tuffites and siltstones which contrast in lithology with the Lower Rhyolitic Tuff Formation, which they conformably overlie.

The best section [7219 5203] is south-west of Dolwyddelan Castle (Figure 19) where individual beds of basic tuffs and tuffites, 0.1 to 1 m thick, are composed of alternations of coarse and fine fragments with varying amounts of carbonate. Grading and cross-lamination are common. Subangular to subrounded lapilli, up to 15 cm in diameter, tend to be concentrated in bands. They consist mainly of devitrified basaltic glass and basalt with less common fragments of chloritised tuffaceous siltstone. The matrix varies from a fine indistinct aggregate of carbonate, chlorite, iron ore and feldspar (E 38728) to a mosaic of platy carbonate (E 38730) (Figure 20). Lapilli of vesiculated basaltic glass are replaced by green chlorite and carbonate and the basalt fragments are almost entirely carbonated. The finer tuffite bands show some resemblance to the acid tuffs, though they generally contain sufficient chlorite in the matrix (E 38731) to indicate their relationship to the coarser, basic tuffs.

Interbedded with the basic tuffs in Afon Ystumiau [7312 5282] are thin flaggy beds of fine-grained acid tuff

Figure 19 Dolwyddelan Castle, built on the Upper Rhyolitic Tuff Formation on the northern limb of the syncline

Figure 20 Photomicrograph of basic pumiceous tuff with a clast of vesicular basaltic glass altered to chlorite, carbonate and iron ore and with a small basalt fragment top right. Bedded Pyroclastic Formation. E 38524, ×40

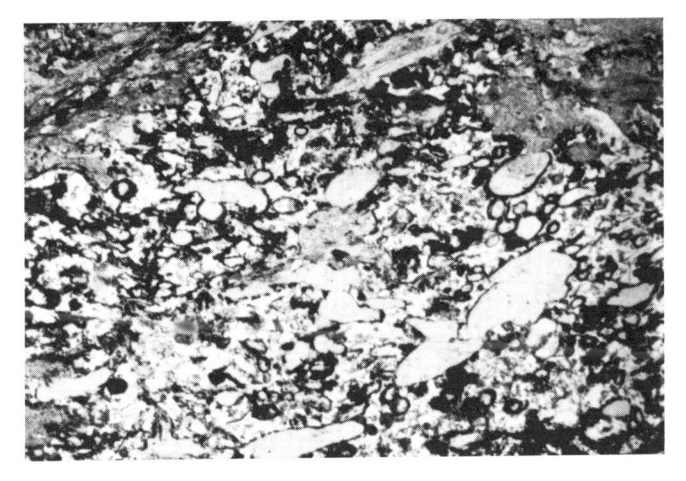

(E 38682). They contain dispersed shardic fragments in a fine siliceous matrix and because they also include even laminae of dark grey mudstone, they probably accumulated by air-fall.

Near Dolwyddelan Castle the formation can be crudely subdivided into a lower division composed mainly of tuffs and an upper division composed mainly of tuffaceous siltstones. The latter are overlain by a fine-grained acid tuff which can be traced eastwards along the north limb of the syncline. This tuff is separated from the Upper Rhyolitic Tuff Formation by approximately 6 m of black slates in which shards and euhedral feldspar crystals are scattered throughout. A similar slate separates the Middle and Upper Crafnant Volcanic formations in the Sarnau area.

A transgressive junction between tuff and sediment is apparent on the north bank of the Afon Lledr, south of Dolwyddelan. Veins of tuff are intruded into adjacent siltstones, and blocks of indurated siltstone are incorporated in the tuff. The tuff, poorly bedded, coarse and agglomeratic, passes westwards into fine bedded tuffs with agglomeratic bands. Here the pyroclastics form the upper part of the formation. Eastwards from Dolwyddelan the formation is predominantly composed of grey siltstones with impersistent layers of basic tuff and tuffaceous sandstone. Small pods of altered basaltic rock, restricted to this horizon, occur around the eastern part of the syncline.

Interpretation The agglomerates and tuff intrusions about the Afon Lledr are indications of local volcanic activity. From the marked westward thickening of the tuffs on the northern limb, however, it may be inferred that the bulk of the basic tuffs were derived from a major eruptive centre west of Dolwyddelan. The generally even bedding and the abrupt grading caused by the layers of coarse pumice fragments at the bases of otherwise uniformly fine-grained tuffs suggests that the material was transported and deposited as ash fall, though cross-lamination in some beds suggests reworking in a marine environment. The associated, evenly bedded, locally fossiliferous sandy siltstones and fine sandstones are consistent with shelf deposition.

Upper Rhyolitic Tuff Formation

Outcrops of the Upper Rhyolitic Tuff Formation, 54 to 75 m thick, on the northern and southern limbs of the tightly folded Dolwyddelan Syncline lie no more than 150 m apart between Ty'n-y-Ddol Quarry and Pont Tan-y-Castell. Dolwyddelan Castle is sited on the northern crop of the formation. East of the castle, between Ty Ucha'r Ffordd and Pen-y-Gelli, faulting and plunge variations around the easterly closure of the structure have greatly enlarged the width of outcrop.

The rocks are poorly bedded to massive, and are typically bluish grey. Lithologies are distinctly heterogeneous and range laterally from clean vitric tuffs to tuffaceous mudstones. Some of the mudstones are disturbed, others contain sporadic pods of pyroclastic material, and isolated crystals of feldspar can be seen in a fine muddy matrix. In general the formation tends to be more muddy towards the eastern closure of the structure. Bedding is laterally impersistent. Cleavage is generally well developed throughout the formation, reflecting the high muddy fraction.

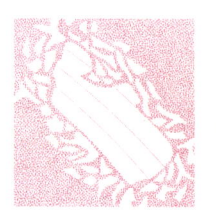

Thin section examination shows that the heterogeneous admixtures of pyroclastic and fine argillaceous material, observed at outcrop extend to the finest elements. Specimens range from predominantly argillaceous (E 38723) through dominantly pyroclastic (E 38724) to clean vitric tuff (E 38727). Shards range in size and shape from fine and elongate, less than 0.1 mm, to angular cusp-edged bubble walls, up to 2.5 mm. Rare crystals of albite-oligoclase, up to 3.8 mm (E 38514), are invariably much altered and resorbed. Clasts of tubular pumice, many with indistinct margins, occur throughout the formation (E 38514).

Cleavage is most prominent around the eastern closure of the syncline, where the planes are accentuated by iron oxide, and possibly opaque carbonaceous material and there is a strong alignment of flakes of secondary mica adjacent to the planes (E 38684).

Resorbed sodic plagioclase

Tubular pumice

Interpretation Williams and Bulman (1931) assumed that the irregular admixture of muddy and pyroclastic material resulted from 'rapid re-assorting and mixing in shallow water', but the lack of bedding does not support such a conclusion. On the contrary, the broad lateral extent of the formation together with the complete lack of sorting suggests deposition from high density turbid flows following the remobilisation of ash previously emplaced on unstable slopes of unlithified mud.

Llanrhychwyn Slates and Black Slates of Dolwyddelan   4

Graptolite
Orthograptus calcaratus
1½ times life size

The Llanrhychwyn Slates are correlated with the Black Slates of Dolwyddelan on the basis of their lithologies and their relationship to the underlying volcanic rocks. The slates are graptolitic and all the faunas collected during the survey have been assigned to the *Diplograptus multidens* Zone (p. 70) which Skevington (1969) advocated to replace the *Climacograptus peltifer* and *C. wilsoni* zones—individually unrecognisable in North Wales. Limitations to the usefulness of the determinations are the predominance of biserial graptoloids, the poor state of preservation and the uncertainty as to the limit of the *D. multidens* Zone, particularly as this zone may overlap with the overlying *Dicranograptus clingani* Zone (Skevington, personal communication, 1971). Thus the present determination of the *D. multidens* Zone and the earlier recognition of the *D. clingani* Zone in the Black Slates of Dolwyddelan (Bulman and Williams, 1931) should not be read together as meaning, necessarily, that both zones are present either wholly or in part within the district.

The problem is further accentuated by a lack of agreement as to the general relationship of the graptolite zones to the stages based on shelly faunas. Both faunas occur within black argillaceous sediments with phosphatic nodules (the Nod Glas horizon) in the Welshpool area, where Cave (1965) assigned the graptolites to the *D. clingani* Zone and the shelly forms to the Actonian and Onnian stages, the Actonian resting nonsequentially on the Longvillian. On that basis, Cave's further correlation of the Nod Glas with the Black Slates of Dolwyddelan would imply that the latter also represent the Actonian and Onnian stages. Apart from the questionable precision of the *D. clingani* faunas, however, there are no known nonsequences or shelly faunas in the Black Slates of the Capel Curig and Betws-y-Coed district.

Llanrhychwyn Slates

The junction between the Llanrhychwyn Slates and the underlying Crafnant Volcanic Group is not exposed in the two areas where the slates crop out. The largest area is elongated from NE to SW along the core of a syncline extending from Ty'n-y-Groes to Llyn Goddionduon. Exposures there are numerous, but they are generally small, except in old quarries, as on the south side of Coed Tal-y-Llyn, where about 35 m of thickly bedded, well cleaved bluish black mudstone with hard iron-rich bands are exposed.

In the other area of outcrop, east of the Llyn-y-Parc Fault, exposures in a stream section near Aber-llyn Mine show the slates to be brecciated and mineralised with associated sulphurous weathering. Farther north the outcrop is largely obscured by scree, but a roadside exposure [8000 5854], 0.5 km N of Cwmlanerch, yielded graptolites indicating a *D. multidens* Zone age (p. 70).

Black Slates of Dolwyddelan

The Black Slates are black, 'sooty' and graptolitic with pyrite along bedding and cleavage. They overlie the Snowdon Volcanic Group forming the core of the Dolwyddelan Syncline. Within the district they are exposed between Tan-y-Castell in the east and Ty'n-y-Ddol Quarry in the west, and have been quarried extensively in the past at Chwarel Ddu and Ty'n-y-Ddol. The junction with the underlying volcanic rocks is exposed only in the north side of Chwarel Ddu where it shows evidence of slipping or thrusting.

Bulman and Williams (1931) assigned graptolites from the Black Slates to the *D. clingani* Zone. Faunas collected during the present survey indicate the *D. multidens* Zone, but after examining Bulman's material Professor Skevington and Dr R. B. Rickards confirmed that it represents a *clingani* Zone assemblage.

Interpretation Williams and Bulman (1931) assumed that the Black Slates of Dolwyddelan were deposited in shallow water because they conformably overlie the supposed shallow-water admixed lithologies of the 'Upper Rhyolite Series'. They accepted the graptolites as evidence of water circulation, taking the restricted fauna, the occurrence of pyrite and the black colour of the slates to indicate a lack of oxygen and circulation where the muds were deposited. Cave (1965) assumed that the Nod Glas sediments, the presumed correlatives of the Black Slates, were similarly deposited in shallow water covering a rising shelf. He thus explained the non-sequences between the shallow-water Caradoc sediments below and Ashgill sediments above the Nod Glas as well as the phosphorite that forms a conspicuous part of a condensed Nod Glas sequence in the Bala –north Montgomeryshire area. Such an uplifted shelf implies gradients not only to the deeper water basin which lay to the east and which was receiving clastics from the east, but also to deeper water in the west and north, in the area of the Capel Curig and Betws-y-Coed district.

Although contrary to the view of Bulman and Williams (1931), this deeper water interpretation for the Black Slates of Dolwyddelan is here preferred. It explains the upward passage from the coarse terrigenes of the Carneddau Group to the

Extent of Grinllwm Slates

Extent of Llanrhychwyn Slates and Black Slates of Dolwyddelan black slates of Middle Crafnant and subsequent formations. Moreover, it is improbable that such an uplifted landmass to the north would have been exhausted as a source of coarse sediments in a volcanically active area and that a wide area of North Wales was fortuitously shielded from that source. It is more likely that the landmass was progressively submerged and that the depositional basin was covered by deeper water than hitherto in the Caradoc.

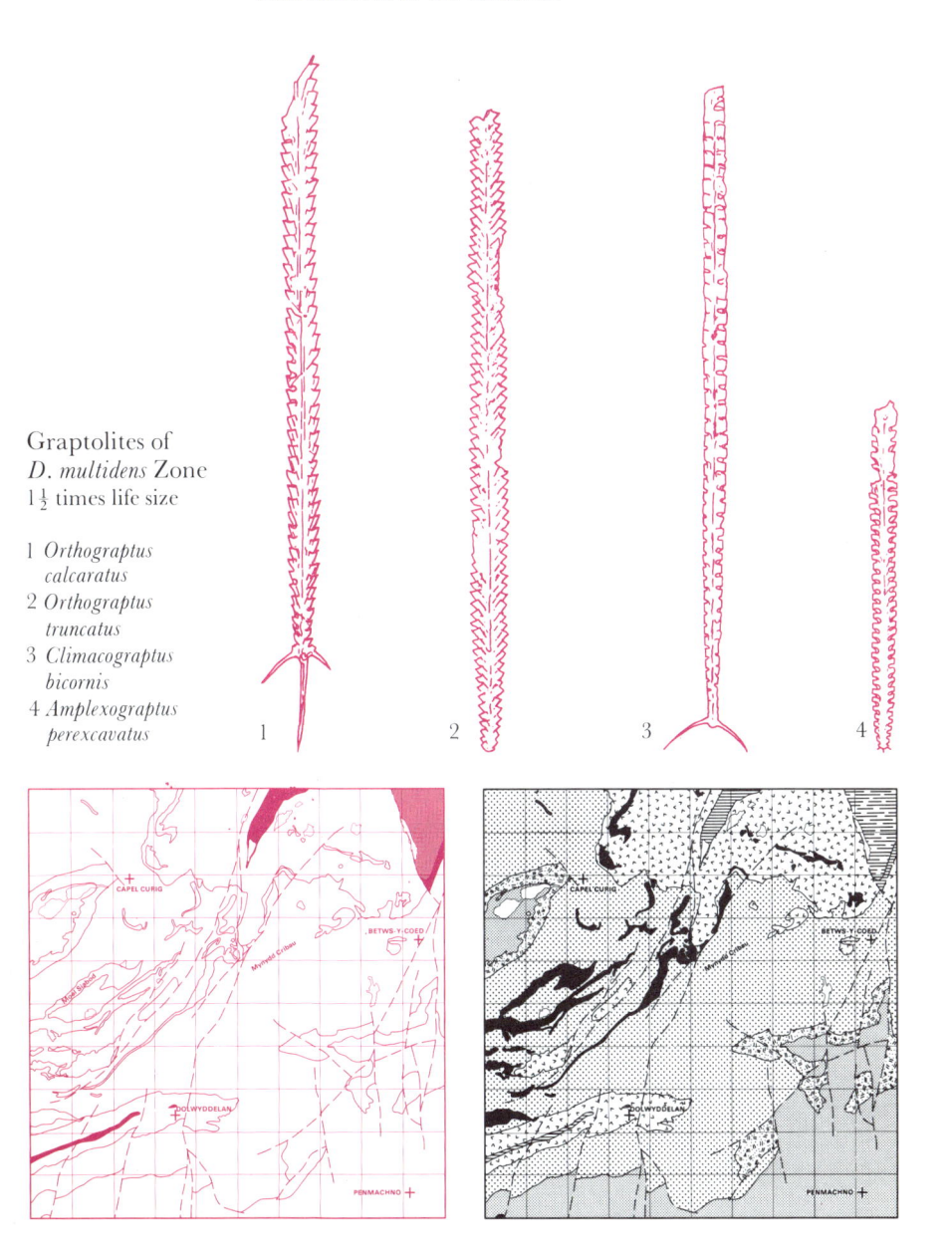

Graptolites of
D. multidens Zone
1½ times life size

1 *Orthograptus calcaratus*
2 *Orthograptus truncatus*
3 *Climacograptus bicornis*
4 *Amplexograptus perexcavatus*

5 Grinllwm Slates

The Grinllwm Slates are exposed only in the faulted outlier around Gwydir High Park, in the north-eastern corner of the district. There they conformably overlie the Llanrhychwyn Slates, though farther north, in the Trefriw district, they are separated from that formation by basaltic tuffs and associated mudstones (Davies, 1936). They form a prominent scarp on the west side of the Conway Valley south of Llanrwst, where they are estimated to be 220 m thick.

The beds are cleaved grey silty mudstones with irregular thin bands and lenses of sandstone. The bands, generally less than 4 cm thick, are ripple-marked and cross-laminated, medium- to coarse-grained and commonly show soft iron-stained weathering products. There are also rare ironstone nodules. Trilobite fragments have been obtained at one locality [7963 5992]. Along the Llyn-y-Parc Fault the beds are brecciated and mineralised.

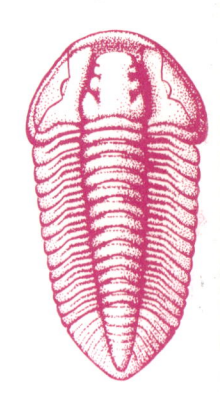

Trilobite
Flexicalymene caractaci
Life size

Dolerites 6

Ophitic texture in dolerite

Figure 21 View north-eastwards from the summit of Moel Siâbod with dolerite forming the crags in the foreground, the Llugwy Valley in the middle distance and Denbighshire in the far distance

Intrusions of dolerite are most prominent in the northern and western parts of the district, where they cut strata ranging from Carneddau Group slates overlying the Capel Curig Volcanic Formation up to the Middle Crafnant Volcanic Formation. Most of them are in the form of sills which are locally transgressive, but there are also a few small bodies of indeterminate morphology, together with a few dykes trending in various directions.

The largest intrusion is bifurcated along the spine of Moel Siâbod to give two sills, each about 150 m thick, forming cliffs above and ridges below the cwm of Llyn-y-Foel. The southernmost sill, itself locally in two leaves, occupies an horizon within the sandstone-tuffite sequence of the Carneddau Group. It can be traced almost continuously within that sequence north-eastwards along the flank of a syncline to the Afon Llugwy (Figure 21), then back south-westwards on the opposite flank of the structure. Such distribution indicates emplacement before the main folding—a view supported by the presence of cleavage, though this tends to be local here and in other intrusions. Most intrusions clearly predate the faulting, but there is equivocal evidence to suggest that others are later than, or are controlled by, faulting, as exemplified by the intrusion south of Craig Forris [757 570]. Thus it would appear that there was more than one period of either intrusion or of faulting.

At outcrop the dolerites are generally greyish green rocks with a brown massive rounded 'granitic' type of weathering. They tend to be more resistant to erosion than the sediments and thus form crags, though in places, and particularly at their margins, they are marked by peaty depressions. At many margins columnar jointing is common. Even where the sills appear from their outcrop pattern to be generally concordant with adjacent sediments and tuffs the marginal contacts are often aberrant. On the southern slopes of Moel Siâbod, for instance, well exposed upper and lower contacts change inclination within a few hundred metres along the strike, from $50°-80°$ to the SE, concordant with the sediments, to similar but discordant angles towards the north-west.

The extent of contact alteration appears to bear little relation to the thicknesses of the intrusions. The widest observed zone of induration and spotting is about 10 m, flanking the sill immediately to the north-east of the summit of Moel Siâbod, but this is exceptional. Similarly, dolerite textures bear little relationship to the thicknesses of intrusions. Traced inwards from narrow zones of chilling at the margins, most rocks are medium-grained, with no apparent progressive increase in coarseness. Ill-defined coarser leucocratic pockets with bladed ferromagnesian minerals may be scattered throughout, as in the sill below Llyn-y-Foel. One of the coarsest rocks, cropping out east of Caer Llugwy [749 524] is gabbroic almost up to the margins, though the sill is not one of the thickest in the district.

Microscopic examination suggests that the dolerites were probably composed originally of clinopyroxene (commonly titaniferous augite), zoned labradorite and opaque titaniferous oxides. In some specimens (E 35467) pseudomorphs after olivine are also prominent and in others accessory apatite is conspicuous (E 38149). Textures are commonly ophitic, with feldspar prisms up to 1 mm long and 0.2 mm across (Figure 22). In the coarse, gabbroic variants (E 35468) the feldspars are as large as 4.5 × 2.5 mm. Some rocks contain ocelli filled, wholly or in part, by finer grained dolerite.

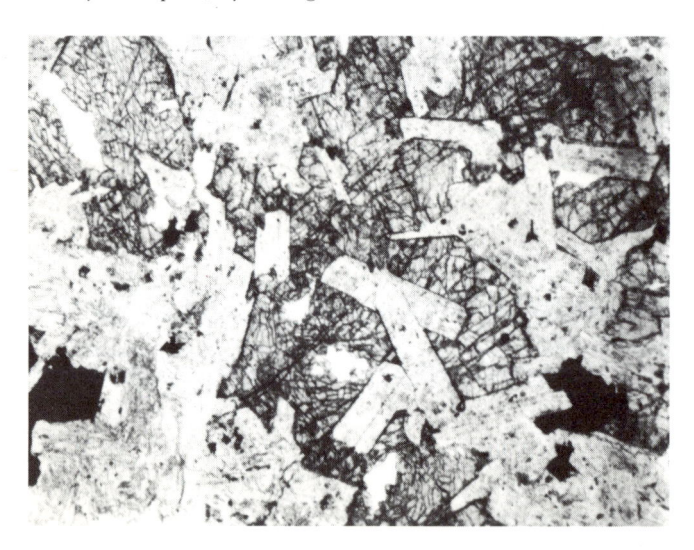

Figure 22 Photomicrograph of dolerite with ophitic texture, consisting of augite, plagioclase and iron ore. E 35194, ×40

Extent of the dolerites

All the intrusions have suffered severe deuteric alteration and some have been further modified by regional metamorphism. The most obvious deuteric effect is the total non-destructive albitisation of original labradorite to oligoclase-andesine. The original labradorite is inferred from the texture and by analogy with specimens of the same suite to the north of the district, where relict calcic plagioclase is preserved. Other deuteric effects include the growth within plagioclase, or outwards from pyroxene, of bladed to fibrous or even hair-like sheaves of actinolite; the development of leucoxene from ilmenite; and, in the final stages, the growth of chlorite and epidote followed by quartz and carbonate in intersertal spaces and ocelli.

The effects of subsequent regional metamorphism are most easily recognised in the imprint of a tectonic foliation. This is clearly indicated by a further oriented growth of chlorite, quartz and carbonate at the expense of feldspar and pyroxene, the latter being typically totally replaced. Dolerites free from pyroxene commonly contain poikiloblastic, graphic, strained quartz crystals, probably dating from the regional metamorphic phase of alteration. The first effect of foliation on the feldspars is the development of laminae; lithons of oriented chlorite grow along the laminae until only fragmented remnants of plagioclase remain in a totally modified igneous fabric (E 35262).

Floyd and others (1976) have provided geochemical data on these rocks, which they describe as meta-dolerites, probably formed quite near the margin of a stable continental plate.

Tertiary dykes are reported from Snowdonia (Williams and Ramsay, 1959) and there are petrographical grounds for ascribing a Tertiary age to a dyke proved underground in a borehole in the north-eastern part of the district (Archer and Elliot, 1965). None have been found at outcrop during the survey; however, a 2.5-m dyke on Creigiau'r Garth, some 0.5 km S of Llynnau Mymbyr, has an appropriate ESE trend, but it is cleaved and highly altered.

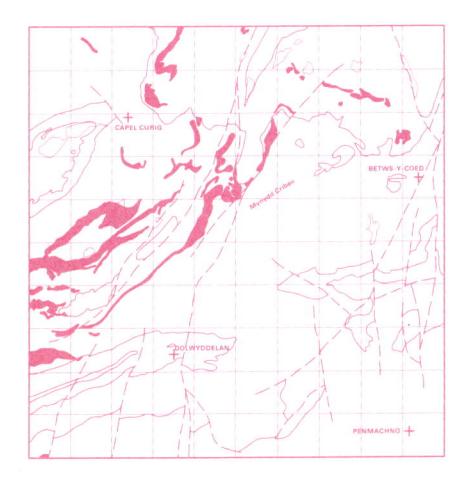

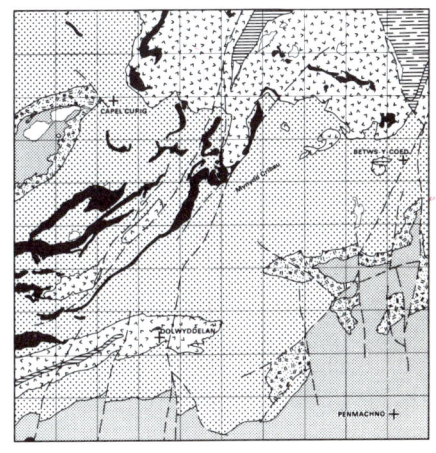

7 Structure

Folds

The district lies within the Snowdon Synclinorium, which has a general north-easterly plunge. The component folds are gentle to isoclinal with axial planes dipping steeply to the north-west and gentle plunges variable in direction. Measurements normal to the bedding in mesoscopic folds show no marked hinge zone thickening. Trends of the major folds vary between north-north-easterly in the northern part of the district and easterly in the south. This variation is illustrated by two of the principal structures, the Capel Curig Anticline and the Dolwyddelan Syncline, which are periclines extending beyond the western margin of the district.

The Capel Curig Anticline, formed of rocks low in the sequence, occurs within, and is congruous with, the central zone of the synclinorium. It typifies the folding in the northern part of the district in being an open structure. Farther east, around Cribau, the folding is mesoscopic and similarly open. To the south, on Moel Siâbod, the folding is again open, although axial-plane traces indicate components of the E–W trend which becomes dominant farther south in the Dolwyddelan Syncline and the complementary Lledr Valley Anticline. Coincident with the change in trend from north-easterly to easterly there is an increase in the scale and degree of closure of the folds.

The Dolwyddelan Syncline is slightly asymmetrical at its eastern end with its axial plane steeply inclined to the north. To the west the northern limb steepens, to become inverted north of Dolwyddelan. On the southern limb in this area dips are steep in the axial region, but decrease to the south. West of Dolwyddelan Castle the inverted limb dips moderately northwards and the fold is here isoclinal. The bedding-cleavage relationship is everywhere congruous with the major structure. Various levels of the Dolwyddelan structure are seen by virtue of faulting and the variable plunge. A constructed profile shows an axial plane steepening at depth. The overturning of the fold at higher levels would appear to represent a separate deformation during which it was considerably tightened with the rotation of both bedding and cleavage.

The convergence of the major fold axes towards the south-west of the district (Figure 23), together with the increasing scale and tightness of folding in that direction, probably reflects location of the area between two major regional

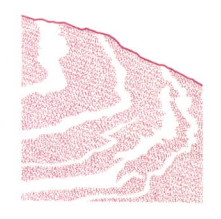

Wavy bedding cut by cleavage

Figure 23 Sketch map showing the distribution and orientation of structures in the district

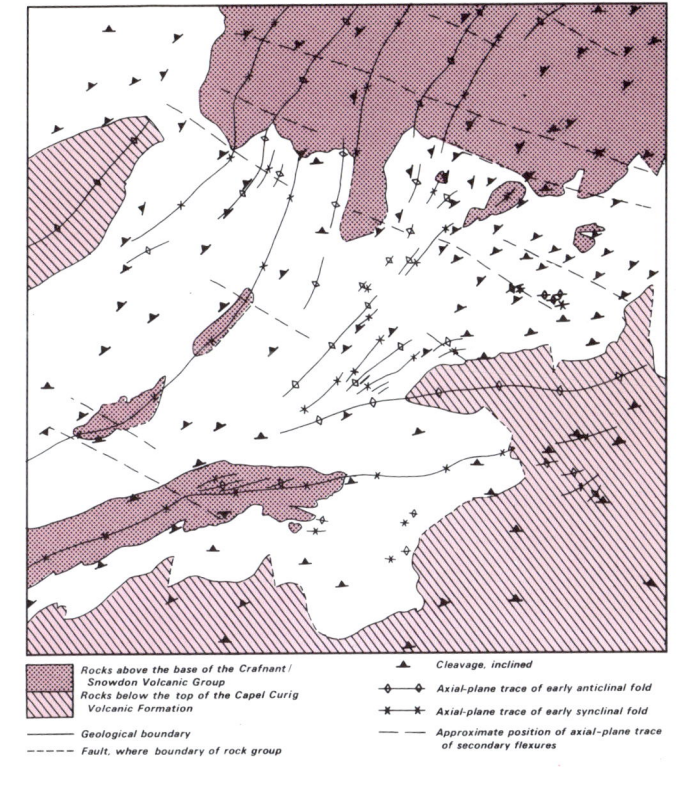

structures—the Harlech Dome and the Snowdon Synclinorium.

The presence of later folds trending between north-west and west-north-west can be inferred from flexures in the major fold traces and associated cleavage. On the other hand, plunge culminations and depressions do not in general appear to reflect a separate phase of deformation and are more likely to be the result of inhomogeneous strain, though this has yet to be demonstrated.

Cleavage

Cleavage related to the axial planes of the folds is generally moderate to steep, though it decreases progressively southwards from Dolwyddelan. Its intensity is closely related to lithology. In mudstone it is slaty, with platy minerals showing a high degree of preferred orientation. In the coarser sandstones of greywacke-type, recrystallisation of quartz and micaceous minerals in the matrix has led to the development of a strong preferred orientation. Intergranular fracturing related to this fabric varies from continuous to discontinuous. Preferred orientation fabrics are less apparent in the finer sandstones, which have a smaller proportion of matrix, and fractures tend to be concentrated into widely spaced zones.

Recent unpublished experimental work (N. Price, personal communication, 1974) suggests that such fractures may be the product of pressure solution.

In the poorly cleaved acid volcanic rocks a tectonic fabric is common in the fine matrix, its intensity being dependent upon the mica content. In highly siliceous tuffs, which lack such a matrix, there is little sign of deformation.

In the slates and tuffites of the Dolwyddelan Syncline mineral growth along cleavage is indicated by elongate pyrite spotting and strain shadows on dispersed tuffaceous inclusions. Elsewhere such growth is ill defined, but clasts, nodules and accretionary lapilli in tuffs, and fossils in sediments, show varying degrees of flattening and extension. The long axes of mineral growth and extension indicators everywhere pitch at a high angle on cleavage planes.

Refraction of cleavage in layers of different composition

Faults

Most of the faults of the district fall into one of two groups, northerly to north-easterly or easterly to east-south-easterly. Faults of the former group, particularly prominent in the east, generally form pronounced features over long distances although many show little displacement. One of the largest is the Conway Valley Fault, which follows the line of the valley south of Waterloo Bridge. Another is the Dolwyddelan Fault which has a northerly trend at Dolwyddelan, but swings to north-easterly and reverses throw farther north before terminating against the Llyn-y-Parc Fault. The trend of the latter does not fit either of the main groups, but it joins the Conway Valley Fault north-east of Betws-y-Coed and is therefore probably contemporaneous with the northerly group. Other faults of this trend bound the block of Upper Crafnant volcanics and Llanrhychwyn Slates around Llyn Goddionduon.

The easterly trending group of faults have a marked effect on the topography in certain areas; one for instance bounds a prominent scarp between Bryn-y-Fawnog and Sarnau. Others partly control the courses of the Afon Llugwy, downstream from Swallow Falls (Figure 24), and the Afon Lledr.

In general the northerly faults terminate against or are displaced by the westerly faults.

Structural history

It is probable that a regional structure approximating to the Snowdon Synclinorium was formed at an early stage in the Caledonian movements. Another early structure has been postulated by Shackleton (*in* Beavon, 1963), who described an arcuate syncline, crossed obliquely by the main Caledonian cleavage, which he interpreted as a rim syncline surrounding a magmatic dome in Central Snowdonia. The concept of a volcanotectonic structure was later developed by Rast (1969) and Bromley (1969), but no entirely satisfactory evidence of

Figure 24
Waterfall at Pont Cyfyng, Capel Curig, where the Afon Llugwy is deeply incised and potholed in cleaved mudstones in the upper part of the Carneddau Group

the existence of the rim syncline has been forthcoming (Fitch, *in* discussion of Bromley, 1969).

The main deformation is the first phase of Helm and others (1963). A proposition by Lynas (1970) that the low-angle main cleavage of the northern part of the Harlech Dome, just south of the district, preceded the main Snowdonian cleavage was refuted by Bromley (1971). The modification of the Dolwyddelan structure may be a reflection of the second phase of Helm and others. The major flexuring of early cleavage about north-westerly trending axes was regarded as a regional phenomenon by Shackleton (1952) and corresponds to the third phase of deformation of Helm and others (1963). Divergence of the major folds from original Caledonian trends is seen as a result of this or even later deformation.

As has been suggested by others (e.g. Roberts, 1967) it would appear that the maximum compression during the main deformation phase was between north-west and south-east. Apart from the Afon Lledr structures, none of the folds of the district have the strong asymmetry which elsewhere in North

Wales (Shackleton, 1952) is taken to indicate relative transport of overlying rocks towards the north-west. The F_3 structures suggest maximum compression at right angles to that of the main deformation phase (Helm and others, 1963).

The dating of mineral deposits in the Llanrwst mining field (p. 55) indicates emplacement during the Carboniferous. This suggests that the occupied faults are pre-Hercynian structures. They post-date cleavage and appear to be undeformed by F_3 folds, suggesting a late Caledonian age. Brecciation of mineral deposits shows that there has been later movement along these planes.

In the Capel Curig Anticline, the parallelism of minor faults and a cleaved dolerite dyke suggests that those faults may be early Caledonian structures.

Faults, where exposed, are seen to be steeply inclined normal faults showing no signs of lateral movement. The north-westerly dyke trend is common in North Wales and is regarded by Shackleton (1952) as indicating emplacement when maximum regional stress was NW–SE, i.e. during the main deformation phase. The fault pattern in general appears to bear no fixed relationship to the main Caledonian structures.

Figure 25 Sketch of abandoned mine near Sarnau

Mineralisation 8

The district includes the southern part of the Llanrwst mining field, which was once one of the most important sources of lead and zinc in the Lower Palaeozoic rocks of North Wales. The field is bounded to the east by the Llyn-y-Parc Fault, to the west by the outcrop of the Upper Crafnant Volcanic Formation around Llyn Bodgynydd, and to the south by the Afon Llugwy. It includes the properties of Aber-llyn, Coedmawr Pool, Ffrith, Gorlan, Cyffty, North Cyffty, Llanrwst and part of Parc and Hafna, the boundaries of which are shown, together with the principal lodes and mines of the district, in Figure 26.

Figure 26 Sketch map showing lodes and properties of the southern part of the Llanrwst mining field

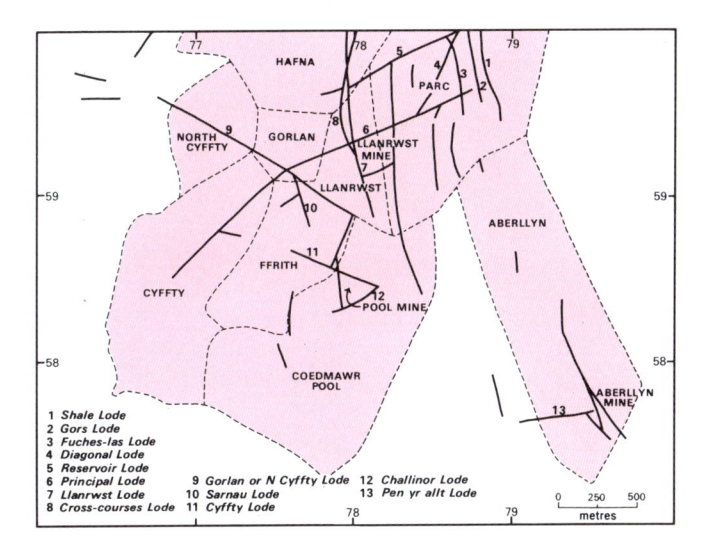

1 Shale Lode
2 Gors Lode
3 Fuches-las Lode
4 Diagonal Lode
5 Reservoir Lode
6 Principal Lode
7 Llanrwst Lode
8 Cross-courses Lode
9 Gorlan or N Cyffty Lode
10 Sarnau Lode
11 Cyffty Lode
12 Challinor Lode
13 Pen yr allt Lode

The mineral potential of the district was recognised as early as 1625 when Sir John Wynne of Gwydir wrote to Sir Hugh Myddleton 'I have leade ore on my ground in great store, and other minerals near my house, yf it please you to come hither' (Dewey *in* Dewey and Smith, 1922, p. 59). Mining activity was most intensive from 1848 to 1914 and was concentrated in the Sarnau area, near the centre of the field. Output during this period amounted to 11 357 tons of lead ore and 12 304 tons of zinc ore. Mining declined as cheaper sources were discovered elsewhere and from 1914 to 1938 the recorded output had dropped to 1501 tons of lead ore and 1424 tons of zinc ore (Figure 25). The Parc Mine lodes, associated in some instances with the Llyn-y-Parc Fault, were worked intermittently until the late nineteen-fifties (Dennison and Varvill, 1952; Archer, 1959).

The following summary is based mainly on publications by Dewey (*in* Dewey and Smith, 1922) and Archer (1959), supplemented by an unpublished report by T. Robertson in 1940, and research by Marengwa (1973).

The mineralisation is of lode type occupying steeply dipping normal faults which generally show little displacement. The lodes have three trends, the earliest east-north-easterly set being displaced by northerly lodes and both displaced by east-south-easterly lodes. The northerly lodes form mineralised belts, up to 80 ft wide, which 'merge insensibly into the country rock on one or both sides' (Dewey and Smith, 1922, p. 60), whereas the other lodes, up to 6 ft wide, have well-defined walls.

Substantial tonnages of ore have been gained from the northerly lodes at the Aber-llyn and Parc mines and from the east-south-easterly lodes in the Pool and Hafna mines. The east-north-easterly lodes were worked principally at Parc, Pool, Llanrwst and Cyffty mines. The nature and disposition of the lodes varies with differing lithologies of the host rock. The inclination of the Principal Lode at Parc Mine flattens to about 45° in black mudstone from the usual dip of about 75° in the more competent tuffs and tuffite. Within the black mudstones the lode is ill defined and can only be traced by poorly mineralised stringers and fault gouge.

The predominant ore minerals are galena and sphalerite, although pyrite and marcasite are locally common. Chalcopyrite and magnetite have also been recorded, but they are rare. The relative abundance of the main gangue minerals, quartz and calcite, is related to their mineral associations and to the wall-rock lithology: in general, breccias of slate are cemented by quartz, and breccias of tuff are cemented by calcite.

Over the mining field as a whole it has been estimated (Dewey and Smith, 1922) that the ratio of ore to gangue averages from 8 to 15 per cent of galena and 8 to 10 per cent of sphalerite. The proportion of galena to sphalerite varies in different lodes; at the Parc Mine the ratio is 2 to 1. Pool, Llanrwst and Cyffty mines produced lead almost exclusively, although zinc is not absent from the lodes. Parc Mine has produced more lead than zinc and Aber-llyn was almost entirely a zinc producer. These variations appear to be local rather than regional, for the distribution both of ores and gangue minerals shows no evidence of mineral zoning.

In virtually all the most important mines dolerite occurs in the vicinity of the lodes, sometimes forming the walls. The

Figure 27 Llyn Elsi south-west of Betws-y-Coed

intrusions are clearly older than the lodes, but it is uncertain whether they have exerted any control on the mineralisation.

Hydrothermal alteration of the wall rock adjacent to the lodes is generally confined to narrow bleached zones, up to 1 m thick, and suggests temperatures of 200°–300°C during mineralisation (Marengwa, 1973). The dominant alteration is silicification, though sericitisation can also be seen in argillaceous sediments. Feldspathisation is patchy in the contact rocks, within 0.3 m of the edge of the lodes, and lead and zinc concentrations associated with this alteration are the highest determined in the wall rocks.

The age of the mineralisation was formerly assumed to be post-Carboniferous (Dewey and Smith, 1922; Archer, 1959), although a dating of 340 ± 70 million years by Moorbath (1962) raised the possibility of an earlier (?Carboniferous) mineralisation. Ineson and Mitchell (1975), using potassium-argon methods, have determined two Carboniferous ages of mineralisation at 320 and 280 million years.

9 Pleistocene and Recent Deposits

During the Pleistocene, North Wales suffered severe glaciation, the complex history of which is still not fully elucidated (Table 1; Whittow and Ball, 1970). Its main effect in the Capel Curig–Betws-y-Coed district was of intense erosion. Glacial drift deposits are relatively thin, of restricted extent and are entirely the products of local ice during the last glaciation (Devensian, Würm or Weischel).

The movement of the ice and its drainage was in a general easterly to north-easterly direction, towards the Conway Valley. A planation between 274 and 305 m on the eastern side of the district is also recognised east of the Conway Valley and indicates that the main ice overrode the valley from the west. For this reason, Warren and others (in preparation) maintain that the valley exerted only local control over an integral North Welsh ice-sheet, contrary to Embleton's (1961) postulate of a Conway Glacier. However, the profiles of the Llugwy and Lledr valleys suggest that in part they were overdeepened by subglacial drainage rather than glacial scouring. Moreover, where the valleys are clearly glacially scoured, as immediately west of Betws-y-Coed and in Glyn Lledr, good hanging-valley features from the south-west are preserved at Rhiwddolion and Gwibernant respectively. The proximity of these features to the Conway Valley would support the existence of a Conway Glacier, albeit at a late stage. The only high-level cwm in the district is that on Moel Siâbod (872 m OD), facing east and occupied by a lake—Llyn-y-Foel (535 m OD). Most of the exposed crags show clear evidence of glacial scouring. The broken ridges of volcanic rocks, on the northern limb of the Dolwyddelan Syncline and the prominent feature of the Garth Tuff at Plas-y-Brenin show good examples of roches moutonnées.

Boulder clay is widespread although it forms moderately extensive tracts only near Dol-Llech in the Llugwy Valley, on the smooth slopes between 700 and 900 ft OD N of Dolwyddelan and in the valley of the Afon Machno. Elsewhere it locally fills hollows, as in the valley about Rhiwddolion. It consists typically of grey to greyish brown tenacious clay with a high proportion of pebbles and boulders, all of local origin.

Fluvioglacial sand and gravel deposits are preserved only as small remnants of high terraces, near Plasglasgwm, near Penmachno, and in the Lledr Valley, near Gethin's Bridge.

Table 1
Glacial deposits of North Wales

Flandrian	Post Glacial	Zones IV–VIII Diatomaceous mud, Cors Geuallt
Devensian (Late)	Late Glacial	Zone III Solifluction, periglacial scree Zone II (Allerød), Cors Geuallt Zone Ic Solifluction, periglacial scree Zone Ib (Bølling), Cors Geuallt
		Welsh local till, Capel Curig

During the period of deglaciation meltwaters were periodically dammed in shallow lakes, many of which (such as Llynnau Mymbyr and Llyn Goddionduon) have survived naturally to the present day. Others were re-established by man during the 19th century and utilised as sources of power, as settling pools in the mining area about Sarnau, and for water supply, as at Llyn Elsi. All that now remains of some of these lakes, however, are marshy peat tracts or mires, as at Cors Geuallt, north of Bryn-tyrch-uchaf. Diatomaceous muds have been recorded at Llynnau Mymbyr, Cors Geuallt, Llyn Sarnau and Llyn Elsi (Thomas, 1972) (Figure 27). At Llyn Sarnau, peat up to 2.5 m with reed and wood remains forms a layer over most of the lake bed, covering a saucer-shaped deposit of peaty diatomaceous mud, generally about 1 m thick, which in turn overlies fine grey glacial clay with slate fragments. Crabtree (1966) examined the deposits at Cors Geuallt and found evidence for both the Allerød Phase and an earlier warm phase which may possibly represent the Bølling Oscillation.

Exposed slopes were subject to solifluction in the Late-Glacial and Post-Glacial stages. Most have a mantle of ochreous weathered clay with angular to subangular rock fragments and isolated well-rounded boulders. In Cwm Penamnen, it is possible to limit the deposits below the rock scarps on the steep valley slopes and above the alluvium.

The screes of the district were mainly formed in periglacial conditions during the Late-Glacial Period. It is only on the higher ground on Moel Siâbod, above the cwm of Llyn-y-Foel, that they have continued to accrue to the present day. On lower ground, they are stabilised and so mixed with hillwash and overgrown that they cannot be delineated; a scree below the east-facing scarp of Grinllwm Slates in Coed-yr-Allt Goch, north of Betws-y-Coed (Ball, 1966) is an example which was dense forest at the time of mapping.

The main alluvial tract extends along the River Conway, about Betws-y-Coed. It includes well-defined terraces 3 and 6 m above the river, the higher probably having been deposited by the Afon Llugwy which joins the Conway, 1 km NNE of the station. The alluvium, commonly exposed in the eroded banks of the river, consists of clayey sand overlying coarse gravels. Smaller terraced alluvial tracts occur near Penmachno and around Afon Lledr near Dolwyddelan village. At Blaenau Dolwyddelan the low-angled profile of a cone, encroaching on the flat, suggests that the flat was lacustrine at some time. Waterfalls or rapids occur in both the Llugwy and Lledr valleys. In the Llugwy Valley these nick points are encountered at Pont-y-Pair (30.5 m OD), near Miners' Bridge (91 m OD), Swallow Falls (122 m OD) (Figure 28) and Cyfyng Falls (183 m OD). Similarly four nick points can be determined in Afon Lledr between Pont-y-Pair and Glan-y-Wern though at slightly different levels. Before they were breached these obstructions would have held the meltwater in small lakes and it is possible that the alluvial flats now seen on their upstream side were, in part, lacustrine.

Figure 28
Swallow Falls near Betws-y-Coed

References

ARCHER, A. A. 1959. The distribution of non-ferrous ores in the Lower Palaeozoic Rocks of North Wales. Pp. 259–276 in *Future of non-ferrous mining in Gt. Britain and Ireland*. (London: Institution of Mining and Metallurgy.)
— and ELLIOT, R. W. 1965. The occurrence of olivine-dolerite dykes near Llanrwst, North Wales. *Bull. Geol. Surv. G.B.*, No. 23, pp. 145–152.
BALL, D. F. 1966. Late-glacial scree in Wales. *Biul. Peryglacjalny*, No. 15, pp. 151–163.
BASSETT, D. A. 1972. Wales. *In* A correlation of Ordovician rocks in the British Isles. *Spec. Rep. Geol. Soc. London*, No. 3, pp. 14–38.
— WHITTINGTON, H. B. and WILLIAMS, A. 1966. The stratigraphy of the Bala district, Merionethshire. *Q. J. Geol. Soc. London*, Vol. 122, pp. 219–271.
BEAVON, R. V. 1963. The succession and structure east of the Glaslyn River, North Wales. *Q. J. Geol. Soc. London*, Vol. 119, pp. 479–512.
— FITCH, F. J. and RAST, N. 1961. Nomenclature and diagnostic characters of ignimbrites with reference to Snowdonia. *Liverpool Manchester Geol. J.*, Vol. 2, pp. 600–610.
BRENCHLEY, P. J. 1964. Ordovician ignimbrites in the Berwyn Hills, North Wales. *Geol. J.*, Vol. 4, Pt. 1, pp. 43–54.
— 1969. The relationship between Caradocian volcanicity and sedimentation in North Wales. Pp. 181–199 in *The Pre-Cambrian and Lower Palaeozoic rocks of Wales*. WOOD, A. (Editor). (Cardiff: University of Wales Press.)
BROMLEY, A. V. 1969. Acid plutonic igneous activity in the Ordovician of North Wales. Pp. 387–408 in *The Pre-Cambrian and Lower Palaeozoic rocks of Wales*. WOOD, A. (Editor). (Cardiff: University of Wales Press.)
— 1971. Phases of deformation in North Wales. *Geol. Mag.*, Vol. 108, pp. 548–550.
CAVE, R. 1965. The Nod Glas sediments of Caradoc age in North Wales. *Geol. J.*, Vol 4, Pt. 2, pp. 279–298.
CLARKE, J. W. and HUGHES, T.McK. 1890. *Life and Letters of the Rev. Adam Sedgwick*. (Cambridge.)
CRABTREE, K. 1966. Later Quaternary deposits near Capel Curig, North Wales. Unpublished PhD thesis, University of Bristol.
DAVIES, D. A. B. 1936. Ordovician rocks of the Trefriw district (North Wales). *Q. J. Geol. Soc. London*, Vol. 92, pp. 62–90.
DENNISON, J. B. and VARVILL, W. W. 1952. Prospecting with the diamond drill for lead-zinc ores in the British Isles. *Trans. Inst. Min. Metall. London*, Vol. 62, pp. 1–21.
DEWEY, H. and SMITH, B. 1922. Lead and zinc ores in the pre-Carboniferous rocks of West Shropshire and North Wales. Part II, North Wales. *Spec. Rep. Miner. Resour., Mem. Geol. Surv. G.B.*, Vol. 23.
DEWEY, J. F. 1969. Evolution of the Appalachian/Caledonian Orogen. *Nature, London*, Vol. 222, pp. 124–129.

DIGGENS, J. N. and ROMANO, M. 1968. The Caradoc Rocks around Llyn Cowlyd, North Wales. *Geol. J.*, Vol. 6, pp. 31–48.

EMBLETON, C. 1961. The geomorphology of the Vale of Conway, with particular reference to its deglaciation. *Trans. Inst. Br. Geogr.*, Vol. 29, pp. 47–70.

FISHER, R. V. 1966. Rocks composed of volcanic fragments and their classification. *Earth Sci. Rev.*, Vol. 1, pp. 287–298.

FISKE, R. S. and MATSUDA, T. 1964. Submarine equivalents of ash flows in the Tokiwa Formation, Japan. *Am. J. Sci.*, Vol. 262, pp. 76–106.

FITTON, J. G. and HUGHES, D. J. 1970. Volcanism and Plate Tectonics in the British Ordovician. *Earth Planet. Sci. Lett.*, Vol. 8, pp. 223–238.

FLOYD, P. A., LEES, G. J. and ROACH, R. A. 1976. Basic intrusions in the Ordovician of North Wales: geochemical data and tectonic setting. *Proc. Geol. Assoc.*, Vol. 87.

FRANCIS, E. H. 1970. Bedding in Scottish (Fifeshire) tuff-pipes and its relevance to maars and calderas. *Bull. Volcanol.*, Vol. 34, pp. 697–712.

— and HOWELLS, M. F. 1973. Transgressive welded ash-flow tuffs among the Ordovician sediments of N.E. Snowdonia. *J. Geol. Soc. London*, Vol. 129, pp. 621–641.

— SMART, J. G. O. and RAISBECK, D. E. 1968 Westphalian volcanism at the horizon of the Black Rake in Derbyshire and Nottinghamshire. *Proc. Yorkshire Geol. Soc.*, Vol. 36, pp. 395–416.

GEORGE, T. N. 1961. North Wales. *Br. Reg. Geol., Inst. Geol. Sci.*

— 1963. Palaeozoic growth of the British Caledonides. Pp. 1–30 in *British Caledonides.* JOHNSON, M. R. W. and STEWART, F. H. (Editors). (Edinburgh and London: Oliver and Boyd.)

HARKER, A. 1889. *The Bala Volcanic Series of Caernarvonshire.* (Cambridge.)

HELM, D. G., ROBERTS, B. and SIMPSON, A. 1963. Polyphase folding in the Caledonides south of the Scottish Highlands. *Nature, London*, Vol. 200, pp. 1060–1062.

HOWELLS, M. F., LEVERIDGE, B. E. and EVANS, C. D. R. 1971. The Lower Crafnant Volcanic Group, eastern Snowdonia. *Proc. Geol. Soc. London*, No. 1664, pp. 284–285.

— — — 1973. Ordovician ash-flow tuffs in eastern Snowdonia. *Rep. Inst. Geol. Sci.*, No. 73/3.

INESON, P. R. and MITCHELL, J. G. 1975. K-Ar isotopic age determinations from some Welsh mineral localities. *Trans. Inst. Min. Metall. London, Sect.B: Appl. Earth Sci.*, Vol. 84, pp. 7–16.

LORENZ, V. 1973. On the formation of maars. *Bull. Volcanol.*, Vol. 37, pp. 183–204.

LYNAS, B. D. T. 1970. Clarification of the Polyphase Deformations of North Wales Palaeozoic Rocks. *Geol. Mag.*, Vol. 108, pp. 548–550.

MACDONALD, G. A. 1972. *Volcanoes.* (Englewood Cliffs: Prentice Hall.)

MARENGWA, B. S. I. 1973. The mineralisation of the Llanrwst area and its relation to mineral zoning in North Wales, with reference to the Halkyn-Minera area and Parys Mountain. Unpublished PhD Thesis, University of Leeds.

MOORBATH, S. 1962. Lead isotope abundance studies on mineral occurrences in the British Isles and their geological significance. *Philos. Trans. R. Soc. London*, Series A, Vol. 254, No. 1042, p. 325.

NATLAND, M. L. and KUENEN, Ph. H. 1951. Sedimentary history of the Ventura Basin, California and the action of turbidity currents. Pp. 76–107 *in* Turbidity Currents and the Transportation of Coarse Sediments to Deep Water. HOUGH, J. L. (Editor). Symposium sponsored by *Society of Economic Palaeontologists and Mineralogists, Tulsa, Oklahoma, U.S.A. Spec. Publ. Soc. Econ. Pal. Mineral.*, No. 2.

OLIVER, R. L. 1954. Welded tuffs in the Borrowdale Volcanic Series, English Lake District, with a note on similar rocks in Wales. *Geol. Mag.*, Vol. 91, pp. 473–483.

RAST, N. 1961. Mid-Ordovician structures in south-western Snowdonia. *Liverpool Manchester Geol. J.*, Vol. 2, pp. 645–652.

— 1969. The relationship between Ordovician structure and volcanicity in Wales. Pp. 305–335 in *The Pre-Cambrian and Lower Palaeozoic rocks of Wales*. WOOD, A. (Editor). (Cardiff: University of Wales Press.)

— BEAVON, R. V. and FITCH, F. J. 1958. Sub-aerial volcanicity in Snowdonia. *Nature, London*, Vol. 181, p. 508.

RAMSAY, A. C. 1866. The Geology of North Wales. *Mem. Geol. Surv. G.B.*, Vol. 3.

— 1881. The Geology of North Wales (2nd Edit.). *Mem. Geol. Surv. G.B.*, Vol. 3.

ROBERTS, B. 1967. Succession and structure in the Llwyd Mawr Syncline, Caernarvonshire, North Wales. *Geol. J.*, Vol. 5, pp. 369–390.

ROBERTSON, T. 1940. *The lead and zinc deposits of Llanrwst district, North Wales*. Unpublished Internal Report, Geol. Surv. G.B.

ROMANO, M. and DIGGENS, J. N. 1969. Longvillian shelly faunas from the Dolwyddelan area, North Wales. *Geol. Mag.*, Vol. 106, pp. 603–606.

SANDERS, J. E. 1965. Primary sedimentary structures formed by turbidity currents and related resedimentation mechanisms. Pp. 192–219 *in* Primary Sedimentary Structures and their Hydrodynamic Interpretation. MIDDLETON, G. V. (Editor). Symposium sponsored by *Society of Economic Palaeontologists and Mineralogists, Tulsa, Oklahoma, U.S.A. Spec. Publ. Soc. Econ. Pal. Mineral.*, No. 12.

SHACKLETON, R. M. 1954. The structural evolution of North Wales. *Liverpool Manchester Geol. J.*, Vol. 1, pp. 261–297.

— 1959. The stratigraphy of the Moel Hebog district between Snowdon and Tremadoc. *Liverpool Manchester Geol. J.*, Vol. 2, pp. 216–252.

SKEVINGTON, D. 1969. The classification of the Ordovician System in Wales. Pp. 161–179 in *The Pre-Cambrian and Lower Palaeozoic Rocks of Wales*. WOOD, A. (Editor). (Cardiff: University of Wales Press.)

STEVENSON, I. P. 1971. The Ordovician rocks of the country between Dwygyfylchi and Dolgarrog, Caernarvonshire. *Proc. Yorkshire. Geol. Soc.*, Vol. 38, pp. 517–547.

THOMAS, D. 1972. Diatomaceous deposits in Snowdonia. *Rep. Inst. Geol. Sci.*, No. 72/5.

TRAVIS, C. B. 1909. On some Ordovician rhyolites and tuffs of Nant Ffrancon, Caernarvonshire. *Proc. Liverpool Geol. Soc.*, Vol. 10, pp. 311–326.

WARREN, P. T. and others. (In preparation). Geology of the country around Rhyl and Denbigh. *Mem. Geol. Surv. G.B.*

WHITTINGTON, H. B. and WILLIAMS, A. 1964. The Ordovician Period. *Q. J. Geol. Soc. London*, Vol. 120S, pp. 241–254.

WHITTOW, J. B. and BALL, D. F. 1970. North-west Wales. Pp. 21–58 in *The Glaciations of Wales and adjoining regions.* LEWIS, C. A. (Editor). (London: Longmans.)

WILLIAMS, A. 1972. *In* A Correlation of Ordovician rocks in the British Isles. *Spec. Rep. Geol. Soc. London*, No. 3.

WILLIAMS, D. 1930. The geology of the country between Nant Peris and Nant Ffrancon (Snowdonia). *Q. J. Geol. Soc. London*, Vol. 86, pp. 191–233.

— and RAMSAY, J. G. 1959. Geology of some Classic British Areas, Snowdonia. *Geol. Assoc. Guide*, No. 28.

WILLIAMS, H. 1922. The igneous rocks of the Capel Curig District (North Wales). *Proc. Liverpool Geol. Soc.*, Vol. 13, pp. 166–202.

— 1927. The Geology of Snowdon (North Wales). *Q. J. Geol. Soc. London*, Vol. 83, pp. 346–431.

— and BULMAN, O. M. B. 1931. The geology of the Dolwyddelan Syncline (North Wales). *Q. J. Geol. Soc. London*, Vol. 87, pp. 425–458.

Excursion itineraries

The district lies within the Snowdonia National Park, but most of the ground is owned privately or by the Forestry Commission. Permission should be obtained for access to any areas away from public footpaths and what is clearly open sheep grazing on the higher ground. In addition, users of the excursion itineraries set out below are strongly recommended to conform to the Code of Conduct for Geology published by the Geologists' Association.

A Capel Curig Volcanic Formation, north of Llynnau Mymbyr (half day)

Route From Plas-y-Brenin take the Glyders Footpath westwards, bearing south to the lowest crags of the Garth Tuff. Climb through succession to the Dyffryn Mymbyr Tuff, then traverse eastwards to see lateral variation in the Racks Tuff before returning to the starting point. Walking distance 2.5 km.

1 *Garth Tuff*
The base is not exposed, and the tuffs forming the lowest crags are massive, cream-white and welded, with a few sodic feldspar crystals and isolated siliceous nodules. They pass gradually upwards through tuffs with crude bedding foliation into reworked, flaggy, cross-bedded tuffs.

2 *Fossiliferous sediments*
An old quarry in fine sandy siltstone with thin clast-rich bands and bands containing brachiopod shells.

3 *Racks Tuff*
On the west bank of the stream the irregular base of this tuff is distorted by a strong cleavage. Patches of siliceous nodules are exposed below the sheepfold, at the top of the feature. Within 150 m to the east, the tuff is wedged out and the mudstones below and above converge to form an uninterrupted sequence, though isolated bodies of tuff occur still farther to the east (e.g. at locality 5 below). The glacially accentuated feature at the top of the Racks Tuff provides a good vantage point for viewing the geological setting of the north-eastern part of the district.

4 *Dyffryn Mymbyr Tuff*
Typically epiclastic tuff and tuffite, with both whole and fragmented accretionary lapilli, are seen in small isolated exposures.

5 *Racks Tuff*
At the north-eastern margin, the contact of this isolated mass of tuff is extremely irregular and fingers into the adjacent sediments. Thin sections show the tuff to be welded up to the contact.

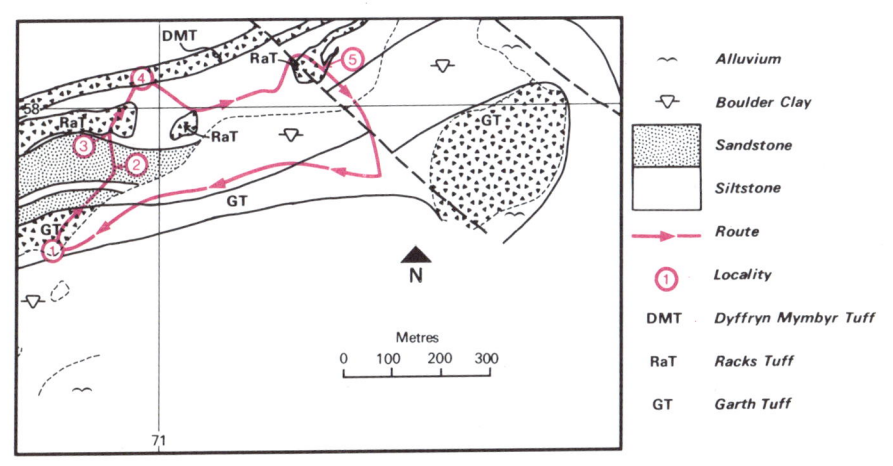

B Carneddau Group and Lower Crafnant Volcanic Formation north-east of Capel Curig (half day)

Route   From the main road, take the path northwards between the Youth Hostel and the Bryntyrch Hotel. From Curig Hill head north-eastwards up the succession to Clogwyn Cigfran: return southwards along track from Nant Geuallt. Walking distance 3 km.

1,2 *Spilitic tuff agglomerate of Curig Hill*
Bedded basic tuffs with isolated basaltic bombs crop out near the path (1). Farther up Curig Hill (2) dips are steeper and a zone of slumping is seen at the centre of the outcrop. On the eastern flanks of the hill the tuffs include a plane of disconformity, and the overlying sediments are inter-fingered with reworked pyroclastic material.

3 *Acid tuffite*
The sandstones above the spilitic rocks contain a thin bed of fine-grained tuffite which has a bleached weathered surface; it has disturbed current-bedding structures at the top.

4 *Mudflow and basic tuff*
This 3-m unit at the top of the sandstones consists of basic tuff below and mudflow breccia containing oriented blocks of sediment and acid tuff above. It is overlain successively by siltstones and mudstones forming the top of the Carneddau Group.

5 *No. 1 Unit*
In this lowest unit of the Lower Crafnant Volcanic Formation lithic clasts are concentrated at the base. The unit shows crude upward grading and is fine and flinty at the top. Clasts of pumice can be distinguished on the weathered surface.

6 *No. 2 Unit*
The outcrop of this unit is reached by crossing first a 'slack' feature formed by poorly exposed siltstones, then a thick dolerite sill. The distinctly chloritic tuff at the base of the unit passes upwards into a clean fine-grained vitric tuff with a few sodic feldspars and distinctive patches of siliceous nodules.

7 *No. 3 Unit*
This unit is characterised by extreme coarseness, an absence of crystals and its general heterogeneity. A band of coarse agglomerate, up to 0.5 m thick, near the top of the unit, is composed of well-rounded blocks with little matrix.

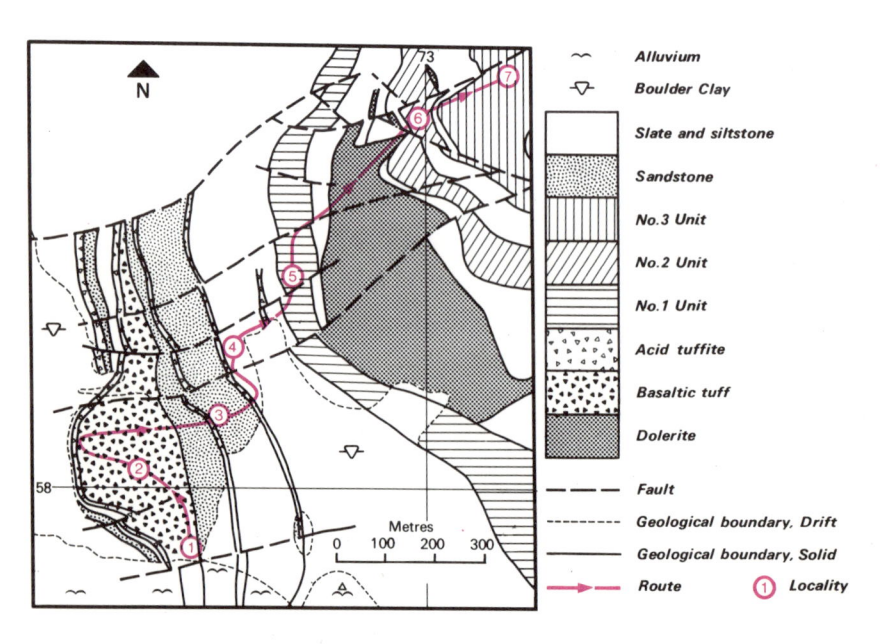

64   EXCURSION ITINERARIES   SH 75

C Lower Rhyolitic Tuff Formation: Moel Siâbod South (whole day)

Route Access is gained by the Forestry Commission Road, leading north off the A496, 1.5 km E of Dolwyddelan, and thence on foot across the peaty alluvial tract adjacent to Afɑn Ystumiau. Walking distance 4 km.

1–3 *Upper Acid Tuff*
Approached from the east, the lower of the two local units of acid tuff is faulted out, and here (1) the crystal-rich base of thick upper acid tuff, with faint bedding lamination, cuts down through basic tuff into sediments. The same acid tuff is again seen on the northern flank of the syncline (2) although the base is not exposed: It shows an upward transition from ill-defined massive bedding with faint lamination, through well-defined massive beds, up to 2 m thick, to well-defined flaggy beds, up to 0.5 m (3). Accompanying this transition is an increase in the epiclastic component and sedimentary structures, including channelling and cross lamination.

4,5 *Lower Acid Tuff*
This unit can be traversed, fairly easily, up the cliffs farther west (4,5). Above the well-bedded crystal-rich base of the unit is a finer tuff which is laminated towards the top. The laminations have been deformed and fragmented by veins and lobate intrusions from the overlying beds. The latter consist of graded silty tuffites which become progressively coarser up the sequence, so that the highest beds are composed entirely of crystal and clast debris (5).

A fine bed of basic tuff is exposed within the fine sediments which separate the lower and upper acid tuffs (between 5 and 6).

6–8 *Upper Acid Tuff*
The coarsely agglomeratic base of the upper tuff is demonstrably channelled by the overlying, thickly bedded vitric tuff. The channelling does not, however, extend down to the basic tuff as seen at locality (1). By descending the crags southwards, across the axial plane of the asymmetric syncline (8), the bedded character of the tuffs can be examined.

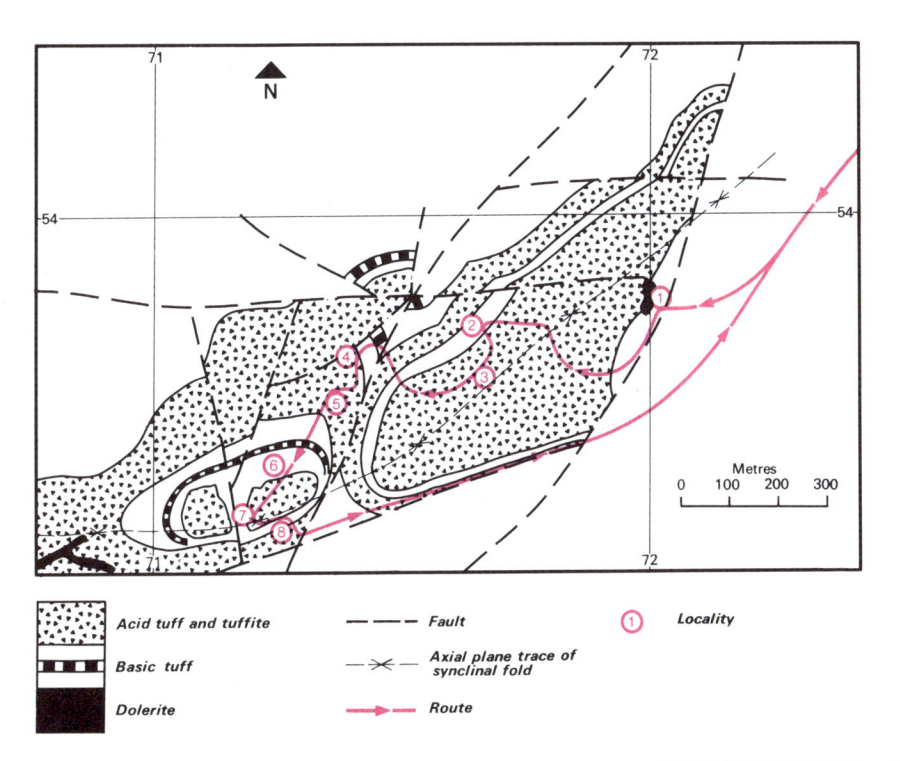

D Capel Curig Volcanic Formation, Pigyn Esgob and Rolwyd (half day)

Route From the road [7754 5145] between Penmachno and Bishop Morgan's Cottage follow the wall westwards to a broad peat depression (1), then walk south-westwards along a scarp of sandstone and siltstone, first to the prominent feature at Pigyn Esgob, then Rolwyd. Return by the same route. Walking distance 3 km.

2 *Pigyn Esgob*

Examine the transgressive character of the main tuff body and of a smaller body farther north. On the southern margin of the main body the original welded fabric is overprinted by perlitic fractures and is almost completely obscured by recrystallisation.

3–5 *Rolwyd*

The irregular base of this welded tuff body transgresses the local sediments (3). Near the contact, here and along the scarp, the sediments are disturbed and are locally totally retextured. In the contact zone the tuff is cleaved and crystal-rich. Traced towards the south-west (4), the irregular base of the tuff forms lobate apophyses into the underlying sediments. A band of mudstone occurs between the contact zone and the main tuff exposed in the higher crags. This band, parallel to the lower contact, curves upwards from nearly horizontal, through vertical in the small gulley at the south-west end of the tuff body (5). The eutaxitic foliation in the main tuff is seen to lie nearly parallel to the regional bedding.

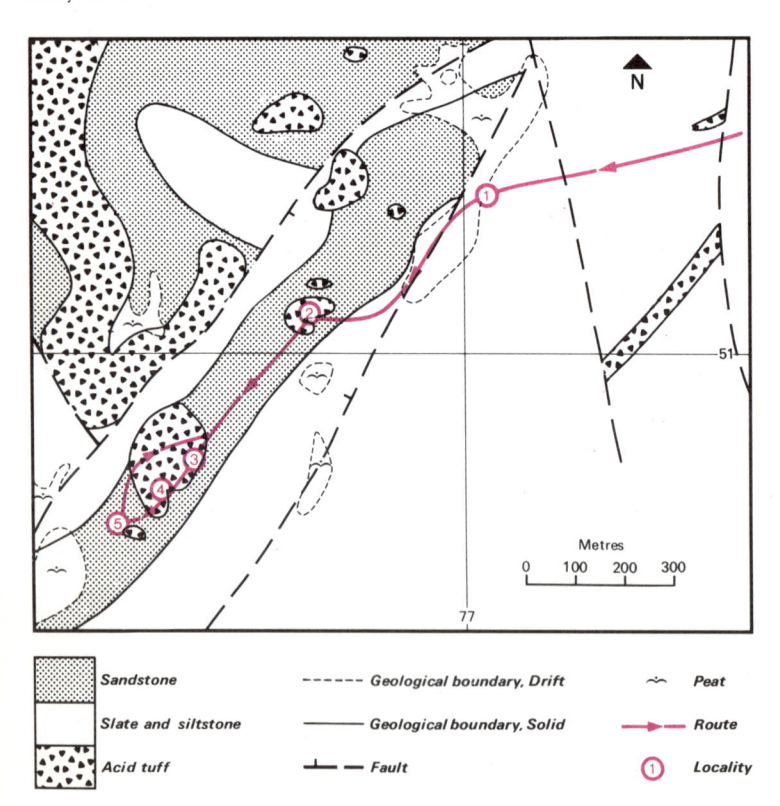

E Snowdon Volcanic Group and overlying Black Slates in the Dolwyddelan Syncline (half day)

Route Take the footpath from the Castle car park across the northern limb of the syncline to the inverted base of the Snowdon Volcanic Group. After inspection of the Lower and Upper Snowdon Rhyolitic tuffs, return to the park. Then take the Blaenau Ffestiniog road to see the Bedded Pyroclastic Formation in the slopes above the road. Access to Chwarel Ddu in the Black Slates is via the slate waste tip. Walking distance 2 km.

1 Lower Rhyolitic Tuff Formation
The lowest of the three tuff units of the formation forms a prominent feature. It is bedded with a central siltstone layer. Clasts are dispersed throughout and there is no apparent grading. Farther south the middle and upper units are variably bedded and show an upward grading, with prominent lithic clasts below and pumiceous clasts above. Scour structures are well exposed on the inverted agglomeratic base of the upper unit and worm-tube casts are present at the top.

2 Upper Rhyolitic Tuff Formation
The formation is well exposed by the path between the Castle and Bryn Seion farm. The lithologies are variable, comprising heterogeneous mixtures of tuff and mudstone. Tuff predominates, forming lenses and more continuous bands within tuffaceous black mudstone.

3 Bedded Pyroclastic Formation
Basic tuffs and tuffites are locally well exposed. They are well bedded with good clastic textures, and many beds are graded with pumice-rich bases.

4 Black Slates of Dolwyddelan
Black slates in the core of the Syncline are exposed in Chwarel Ddu. Bedding and cleavage of the northern limb are locally disrupted by thrusting.

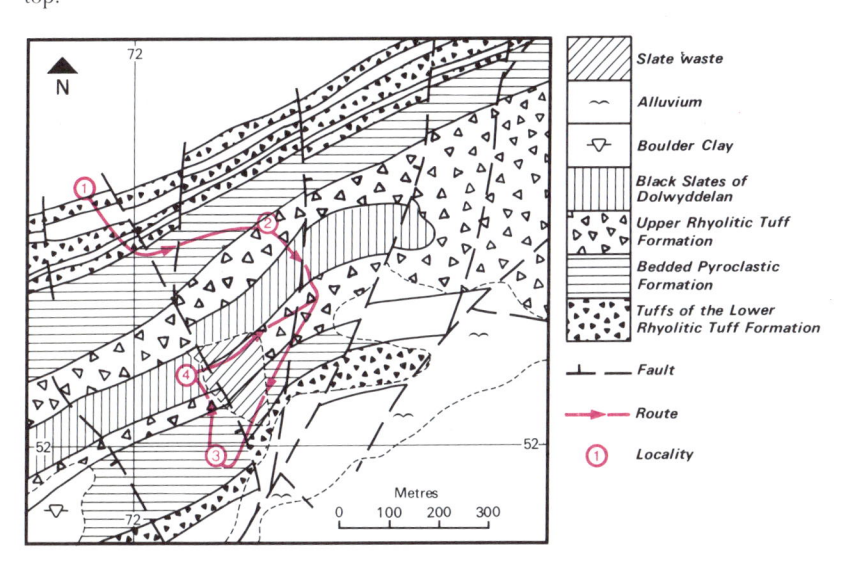

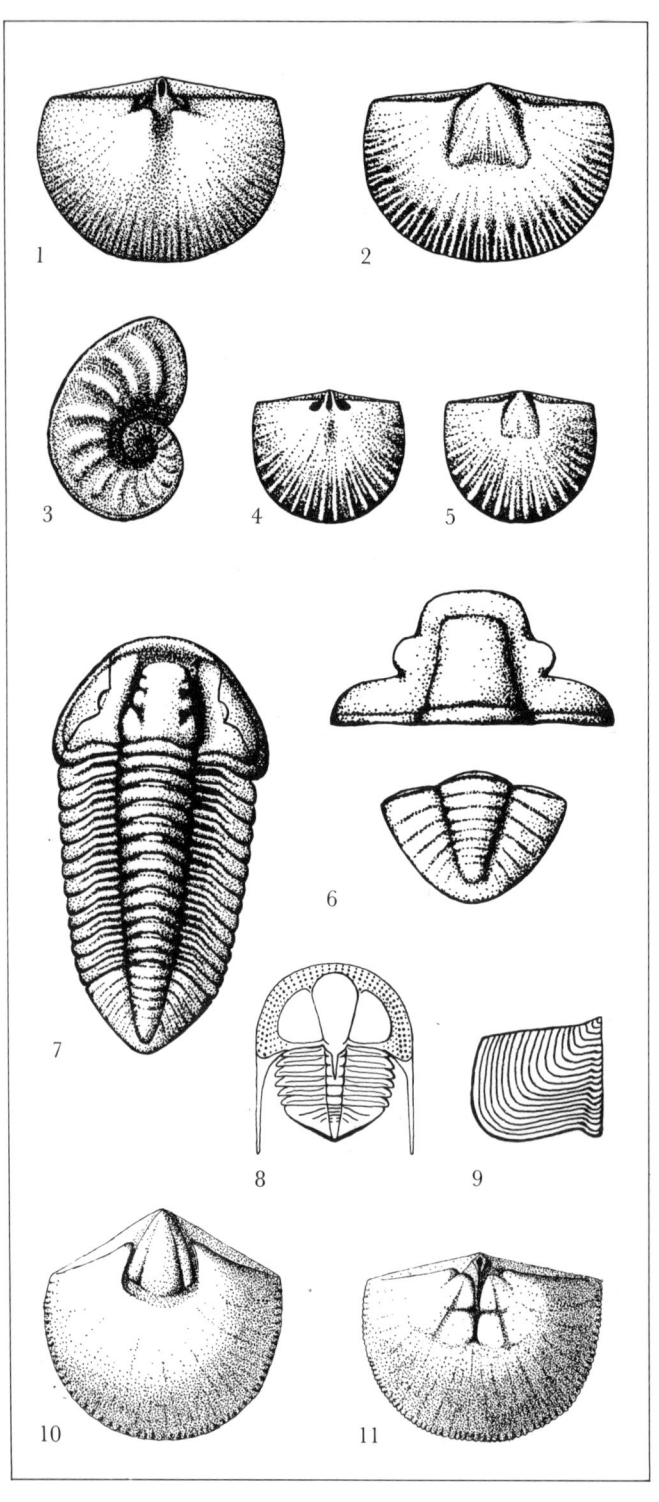

Caradoc shelly fossils

1, 2 *Plaesiomys multifida* (Salter), ×1.
Soudleyan Stage

3 *Cyrtolites nodosus* (Salter), ×1.
Soudleyan Stage

4, 5 *Dinorthis berwynensis* (Whittington), ×1½.
Soudleyan Stage

6 *Brongniartella minor* (Salter), ×2.
Soudleyan and Longvillian stages

7 *Flexicalymene caractaci* (Salter), ×1.
Longvillian Stage

8 *Broeggerolithus nicholsoni (Reed)*, ×1.
Longvillian Stage

9 *Lepidocoleus suecicus* Moberg, a single plate ×4.
Longvillian Stage

10, 11 *Salopia globosa* (Williams), ×2.
Soudleyan Stage

Outside the area of SH 75, some of these species range beyond the stages named here

Fossil localities

The faunas associated with the Ordovician sediments of the district are rarely well preserved and have generally been deformed. Many localities have yielded restricted faunas of uncertain age, but good faunas have been collected from others. A selection of the latter, with lists of determinations are given below. The graptolites were identified by Professor D. Skevington and the other fossils by Dr A. W. A. Rushton

1 [7715 5740] By track 670 m E25°S of Swallow Falls Hotel (RV 3287–3360)
Bryozoa (various forms of colony), *Dalmanella modica* Williams, *Howellites antiquior* (McCoy), *Kjaerina?*, rafinesquinid, *Sowerbyella* cf. *musculosa* Williams, tentaculitid, *Broeggerolithus nicholsoni* (Reed), *Brongniartella minor* (Salter), *Flexicalymene planimarginata* (Reed), *Kloucekia* cf. *apiculata* (McCoy), tetradellid ostracod, crinoid columnals.
This assemblage occurs in the Gelli Grîn Calcareous Ashes of the Bala District (Longvillian Stage), except for *S. musculosa* which is confined to the top of the Allt Ddu Mudstones (upper Soudleyan). The fauna is, however, less diverse.

2 [7865 5759] Crag 180 m S18°E of Pen-yr-Allt Ganol (RV 6249–6256)
Bryozoan, *Cremnorthis sp.*, dalmanellid [juv.], *Dolerorthis sp.*, *Howellites sp.*, *Nicolella sp.*, *Paracraniops sp.*, *Reuschella?*, *Sericoidea sp.*, *Skenidioides sp.* [common], *Sowerbyella sp.*, gastropod and tentaculitid [fragments], *Brongniartella sp.* and calymenid [fragments], *Lepidocoleus sp.* This assemblage probably represents the Longvillian Stage but is not diagnostic.

3 [7852 5755] Crag 225 m S34°E of Pen-yr-Allt Ganol (RV 6257–6267)
Bryozoa, *Cremnorthis parva* Williams [common], *Dalmanella sp.*, *Dolerorthis?* [fragments], *Howellites sp.*, *Nicolella sp.*, *Paracraniops?*, *Reuschella sp.*, *Sericoidea sp.*, *Skenidioides* cf. *costatus* Cooper [common], *Sowerbyella sp.*, *Mimospira?* [sinistral fragments], tentaculitid [fragments], calymenid and trinucleid [fragments], *Lepidocoleus suecicus* Moberg, crinoid columnals [circular].
The species present and the generic constitution of the fauna are reminiscent of the Gelli Grîn Calcareous Ashes of the Bala District, of Longvillian age (Bassett and others, 1966).

4 [7573 5693] Forestry trackside, south of Ty-hyll (RV 6322–6361)
Bryozoa [mainly encrusting], *Dalmanella sp.*, *Dolerorthis duftonensis* (Reed) *prolixa* Williams, *Howellites sp.* [common], *Leptestiina?* [rare], *Nicolella sp.*, *Platystrophia sp.*, *Sericoidea sp.*, *Skenidioides sp.*, *Sowerbyella sp.* [very common], *Modiolopsis* cf. *pyrus* Salter, tentaculitids, *Brongniartella sp.* [fragmentary], *Chasmops* cf. *cambrensis?* Whittington, *Flexicalymene* cf. *caractaci* (Salter), phacopid [fragment], *Cerninella sp.* [common]. *Harperopsis sp.* [common], *Lepidocoleus* cf. *suecicus*, crinoid columnals [pentagonal-rounded and circular].
This assemblage probably represents the Longvillian Stage.

5 [7531 5669] Trackside east of Bryn-y-Gefeiliau (RV 6444–6464)
Bryozoa, *Dalmanella sp.* [small], *Dolerorthis duftonensis* (Reed) [several], *Nicolella sp.*, *Reuschella horderleyensis* Bancroft s.l., *Sericoidea sp.*, *Skenidioides sp.*, *Sowerbyella sp.*, triplesiid?, *Cyclospira?* [fragment], tentaculitid, *Broeggerolithus?* and cybelid [fragments], *Flexicalymene sp.*, *Lepidocoleus sp.*, crinoid columnals [circular].
The fauna suggests the Longvillian Stage. The *Dolerorthis* includes valves resembling both *D. duftonensis* and *D. duftonensis prolixa*, both Longvillian subspecies.

6 [7442 5788] Roadside exposure 400 m N13°W of Dol-gam (RV 6663–6724)
Bryozoa, *Dinorthis sp.*, *Dolerorthis sp.*, *Howellites?*, *Leptestiina* cf. *oepiki* (Whittington), *Nicolella sp.* [abundant], *Reuschella sp.* [abundant], *Sericoidea?*, *Skenidioides sp.*, *Sowerbyella sp.*, *Flexicalymene*

cf. *caractaci* [common], *Platylichas* cf. *nodulosus* (McCoy), trinucleid fragments, *Lepidocoleus suecicus*, primitiid ostracods, tentaculitids, crinoid columnals [pentagonal-rounded and circular].

By comparison with the faunas in the Bala area, a Longvillian age is suggested.

7 [7095 5789] Crags north of Llynnau Mymbyr (RV 6725–6754; 6953–6970)

Bryozoan [stick-like], *Dalmanella sp.* [in siltstone], *Dinorthis* cf. *berwynensis* (Whittington) [in siltstone and sandstone], *Howellites sp.* [in siltstone], *Macrocoelia sp.* [in sandstone], *Salopia* cf. *globosa* (Williams) [in siltstone and sandstone].

The fauna is not diagnostic, but resembles the assemblages at the top of the Allt Ddu Mudstones of Bala (Soudleyan Stage).

8 [7865 5596] Forestry track north of Llyn Elsi (RV 7346–7373)

Bryozoa [stick-like], *Bicuspina spiriferoides* McCoy [several], dalmanellids, *Dinorthis sp.* [costate], *Howellites sp.*, *Leptaena* cf. *ventricosa* Williams, *Rhactorthis?*, *Sowerbyella sp.*, *Broeggerolithus* [sp. indet.], pterygometopid [fragment], machaeridian [fragment], crinoid columnals.

The fauna resembles that at the top of the Allt Ddu Mudstones and in the lower part of the Gelli-grîn Calcareous Ashes at Bala (late Soudleyan or early Longvillian Stage).

9 [7434 5185] Crag on hillside south-east of Dolwyddelan (RV 6971–6986)

Bicuspina sp., *Dalmanella sp.*, *Onniella?*, *Rhactorthis?* [one specimen], *Sowerbyella* cf. *musculosa?*, *Broeggerolithus?* and calymenid [fragments], tetradellid, tentaculitid.

Sowerbyella musculosa is known from the topmost Soudleyan of Bala; *Rhactorthis* is more suggestive of the Longvillian, but is poorly represented here.

10 [7155 5147] Crag north of Afon Lledr 290 m at 75° from Roman Bridge Station (RV 9690–9745).

Bryozoa [stick-like and encrusting], *Dalmanella sp.* [cf. *indica* or *modica*], *Howellites sp.* [indet.], *Sowerbyella* cf. *musculosa* Williams [very abundant],

tentaculitid, *Broeggerolithus?*, calymenid [fragments], *Kloucekia?* [small pygidium], crinoid fragments.

Probably top of the Soudleyan Stage.

11 [7610 5119 to 7610 5102] Crags between Bwlch-y-Groes and Pigyn Esgob (RV 7291–7321)

Massive coarse siltstones yield: *Bicuspina sp.*, dalmanellids, *Macrocoelia sp.* [large], *Salopia globosa* [and another form], asaphoid [fragment] and crinoid columnals.

Cleaved fine siltstones yield: *Bicuspina* [fragment], *Dalmanella sp.*, *Dinorthis* cf. *berwynensis*, *Howellites sp.*, *Macrocoelia* [small], *Cyrtolites nodosus* (Salter), crinoid columnals [round and pentagonal-rounded].

Soudleyan or possibly Longvillian Stage.

12 [7800 5926] By roadside at Llyn Sarnau (RV 3475–3507)

Amplexograptus arctus Elles & Wood, *A. fallax* Bulman, *A.? perexcavatus* (Lapworth), *?Climacograptus antiquus bursifer* Elles & Wood, *?C.brevis* Elles & Wood, *C.lineatus* Elles & Wood, *C.minimus* (Carruthers), *?Glyptograptus euglyphus* (Lapworth), *G.siccatus* Elles & Wood, *Orthograptus* ex gr. *calcaratus* (Lapworth), *O.* ex gr. *truncatus* (Lapworth).

Zone of *Diplograptus multidens*.

13 [8000 5870] Roadside exposure, 0.5 km N of Cwmlanerch (DT 7732–7742)

Amplexograptus sp., *Climacograptus ?brevis*, *C.* cf. *brevis*, *C. lineatus*, *?Glyptograptus euglyphus*, *Orthograptus* ex gr. *calcaratus*, *O.* ex gr. *truncatus*, *Pseudoclimacograptus ?scharenbergi* (Lapworth).

Zone of *Diplograptus multidens*.

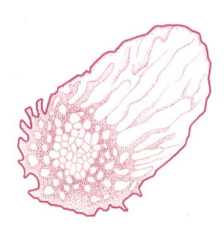

Tabulate coral

Glossary

Accretionary lapilli Pellets formed by the concentric accretion of ash and dust around nuclei of condensed water drops or rock fragments in a volcanic dust cloud
Acid Relating to igneous rocks containing more than 66 per cent of silica
Agglomerate A volcanic rock formed of pyroclastic blocks or fragments generally more than 50 mm diameter
Albitisation The partial or total replacement of the calcic (anorthite) component of plagioclase feldspar by sodic (albite)
Ash flow A turbulent admixture of pyroclastic debris and hot gas which flows in directions imposed by the originating explosive eruption and by gravity
Axial plane The surface that connects the axes of each plane within a fold
Basalt A fine-grained lava or minor intrusion composed mainly of calcic plagioclase and pyroxene with or without olivine
Basic Relating to igneous rocks containing less than 52 per cent of silica
Benioff Zone The plane along which lithospheric plates sink into the upper mantle and where earthquake foci are located
Breccia A coarse-grained clastic rock composed of angular rock fragments set in a finer grained matrix
Caldera A large volcanic depression generally circular in form which may include a vent or vents
Caledonian orogeny Lower Palaeozoic earth movements which reached their culmination at the end of the Silurian
Columnar jointing Prismatic fractures in lavas, sills or dykes which result from cooling
Convolute bedding Complex contorted bedding laminae that are confined to a well-defined undisturbed layer
Cwm An armchair-like hollow generally situated high on the side of a mountain; produced by the downcutting of a glacier
Devitrification The replacement of glassy texture by crystalline texture in a volcanic rock during or after cooling

Disconformity An unconformity in which the bedding planes above and below are essentially parallel
Dolerite A medium-grained igneous rock generally forming minor intrusions and consisting mainly of calcic plagioclase and pyroxene, commonly with an ophitic texture, and sometimes olivine
Epiclastic rock A sedimentary rock formed of fragments derived by weathering and erosion of older rocks
Eutaxitic texture The texture in tuffs where shards and pumice are flattened and deformed around crystal and lithic fragments
Euhedral crystal A crystal showing its natural faces without significant modification
Flame structure Flame-shaped intrusions generally of mud grade that have been squeezed upwards into the overlying generally coarser layer
Fluxoturbidite A sediment deposited under the influence of both turbidity currents and slumping
Gabbro A coarse-grained intrusive igneous rock composed essentially of basic plagioclase and pyroxene with or without olivine
Gangue The uneconomic minerals of an orebody
Greywacke A poorly sorted sandstone with angular to subangular quartz and feldspar fragments and a wide range of lithic fragments set in a clayey matrix
Hyaloclastite A deposit composed of comminuted basaltic glass formed by the fragmentation of the glassy skins of basaltic pillows or by the violent eruption of basaltic material under submarine conditions
Hydrothermal alteration Alteration by or in the presence of water at high temperature
Ignimbrite A form of tuff composed of fragments welded together as they coalesce
Inlier An outcrop of rocks enclosed by younger strata
Isocline A fold with parallel limbs

Lahar A mudflow composed of volcanoclastic material
Lapilli Fragments in the range of 5 to 50 mm ejected by volcanic eruption
Load cast A sole mark composed of sediment of sand grade protruding down into finer grade material and formed as a result of unequal loading
Lode A mineral vein in consolidated rock
Moraine A mound of unsorted debris deposited by a valley glacier (in this account ground moraine is referred to as boulder clay)
Ophitic An igneous texture where prismatic plagioclase crystals are intergrown with pyroxene crystals
Outlier An outcrop of rocks surrounded by older strata
Parataxitic texture An extreme variation of eutaxitic texture in tuffs where the shards are flattened and drawn out
Pericline A fold in which the dip of the beds has a central orientation
Perlitic texture Small-scale arcuate cracks caused by cooling in volcanic glass
Plate tectonics Global tectonics based on an earth model characterised by a number of large lithospheric plates which move on the underlying mantle
Plunge The inclination of a fold axis
Poikiloblastic texture The texture formed where a newer recrystallised mineral surrounds relicts of earlier minerals
Pumice A highly vesiculated glassy lava light enough to float
Pyroclastic A clastic rock formed by explosion or eruption from a volcanic vent
Rhyolite An extrusive igneous rock of acid composition, commonly porphyritic and flow banded
Roche moutonnée An elongate crag scoured by glaciation with a smooth gentle upstream side and a rough steep downstream side
Septarian fractures Radiating fractures at the centres of concretions which intersect concentric fractures and are generally infilled with calcite or quartz
Shard A glass fragment typically found in pyroclastic rocks having distinctive cuspate margins
Sole markings A term commonly used to describe the undersurfaces of a bed infilling underlying sedimentary structures

Solifluction The slow, viscous downhill flow of waterlogged soil or other surface material especially in regions underlain by frozen ground
Spilite An altered basalt in which the feldspar has been albitised and the dark (mafic) minerals altered to low-temperature hydrous minerals
Thixotropy The property of some colloidal substances to change viscosity when sheared; disturbed water-laden sediments may behave in an analogous way
Tuff A lithified deposit of volcanic ash
Tuffite An admixture of pyroclastic (>25 per cent) and epiclastic (>25 per cent) material
Unconformity A break in the stratigraphical sequence marked by a structural discordance
Vent An opening through which volcanic deposits are extruded or ejected
Vesicle A small cavity in a lava formed by included gases
Vitroclastic Texture of a pyroclastic rock composed mainly of cuspate glass fragments
Welded tuff A pyroclastic rock in which individual particles were sufficiently plastic to agglutinate

Index

Accretionary lapilli   11, 13, 71
Agglomerates   30, 35, 71
Air-fall tuffs   6, 16, 37, 39
Albitisation   46, 71
Allt Ddu Mudstones   16, 69, 70
Alluvium   57

Basic tuffs   16, 17, 18, 22, 24, 25, 26, 38
Bedded Pyroclastic Formation   4, 21, 31, 38, 67
Benioff Zone   8, 71
Bodeidda Mudstones   3
Boulder clay   56

Caledonian structures   50, 51
Capel Curig Anticline   15, 48, 52
Carbonate nodules   24, 33, 35
Carreg Alltrem   14
Climacograptus peltifer Zone   41
C. wilsoni Zone   41
Clogwyn Cyrau   23
Clogwyn Manod   30
Columnar joints   14, 45, 71
Contact alteration   45
Convolute lamination   28, 31, 71
Conway Valley Fault   50
Curig Hill   17, 18

Deuteric alteration   46
Diatomaceous mud   57
Dicranograptus clingani Zone   3, 41, 42
Diplograptus multidens Zone   3, 41, 42, 70
Dolwyddelan Castle   39, 40
Dolwyddelan Syncline   20, 21, 33, 38, 40, 48, 50, 51
Dyffryn Mymbyr Tuff   11, 13, 15, 63

Eutaxitic texture   11, 71

Fluviodeltaic sediments   3
Fluxoturbidites   38

Gangue   54, 71
Garth Tuff   11, 13, 56, 63
Gelli-grîn Calcareous Ashes   16, 70
Glanrafon Beds   1, 9, 20
Graptolitic slates   4, 30, 41, 42
Gwydir High Park   44

Hanging valleys   56
Hyaloclastite   6, 22, 25, 26, 35, 71
Hydrothermal alteration   55

Ignimbrite   6

Lledr Valley Tuffs   10, 13
Llyn Bodgynydd   32
Llyn Goddionduon   32, 41, 50
Llyn y Parc Fault   31, 44, 50, 53
Lodes   54, 55, 72
Longvillian   3, 16, 18, 25, 69, 70
Lower Crafnant Volcanic Formation   4, 7, 8, 21, 23, 26, 31, 33, 36
Lower Rhyolitic Tuff Formation   3, 7, 8, 21, 26, 33, 36, 38, 67

Maesgwm Slates   9
Middle Crafnant Volcanic Formation   4, 21, 22, 27, 31, 39, 43, 45
Mines   54

Ophitic texture   46, 72

Parataxitic texture   11, 72
Perlitic fracture   13, 30, 72
Pigyn Esgob   14
Plaesiomys multifida (Salter)   18
Pumice clasts   13, 18, 35, 38, 39, 72

Racks Tuff   11, 13, 63
Rhyolite   6, 8, 20, 30
Rolwyd   14

Sarnau   28
Screes   57
Septarian fractures   14, 72
Siliceous concretions   13, 14, 24, 25
Sills   45
Slate quarries   15, 18, 20, 41, 42
Snowdon Synclinorium   48, 50
Sole markings   27, 72
Solifluction   57, 72
Soudleyan   3, 13, 16, 18, 21, 69, 70
Spherulitic recrystallisation   20
Structural history   50
Sub-glacial drainage   56
Submarine ash flows   7, 11, 31, 36

Tertiary dykes   47
Tuffites   6, 27, 31
Turbid flow   33
Turbidites   27, 31

Upper Crafnant Volcanic Formation   4, 7, 21, 30, 31, 39, 50, 53
Upper Rhyolitic Tuff Formation   4, 7, 21, 39, 40, 67

Welding   6, 7, 11, 13
Welsh Basin   7

Grinllwm Slates

Black Slates (Llanrhychwyn & Dolwyddelan)

Crafnant - Snowdon Volcanic Group

Intrusive dolerite

Carneddau Group

Slates with subordinate sandstones and tuffites

Capel Curig Volcanic Formation

Slates with subordinate sandstones

— — *Fault*

CAPEL CURIG

Moel Siabod